Arsenic
Selenium
Antimony

Ultra-trace chemical analysis
Related notes on bench wet-chemistry and
Spectroscopy techniques.

iUniverse, Inc.
Bloomington

Dr. Paul H. Ramses Gouda
C.Chem., P.R.M.D., Ph.D.

As, Se & Sb analysis:

**Atomic Absorption
Inductively Coupled Plasma
Bench methodology**

The laboratory methodology of ultra trace analysis of As, Se & Sb in environmental, medical & commercial samples.

A university textbook and a laboratory manual.

Front cover by **Pelé Gouda**

Graphic design, cartoon illustration and video animation.

pelegouda@gmail.com

Dr. Gouda's scholastic and professional chemist career, his globally adopted analytical laboratory methods, and his experience as a lead scientist; promise this book to be a reference for every chemist and a must for every university.

*Prof. Dr. Samuel Hein, Ph.D.
Brock University.*

The author, Dr. Paul H. Ramses Gouda, holds two doctorates: MD. with a reseach project in pharmaceutical chemical manipulation of neurochemical compounds, and Ph.D. in ultra trace analysis.

To order additional copies of this book please visit your local branch of any of the following chain bookstores:

- Barnes & Noble
- Chapters, Coles, Indigo

In the absence of shelf copies at your local bookstore, you may still order this book at any "Barnes & Noble" or any "Chapters – Coles" order desk via their in-store computerized database "search & order" kiosk.

This book is also available online at:
www.amazon.com
www.iuniverse.com
www.bn.com   "Barnes & Noble bookstores, USA"
www.chapters.indigo.ca   "Chapters & Coles bookstores, Canada"

iUniverse books may be ordered through booksellers or by contacting:

iUniverse
1663 Liberty Drive.
Bloomington, IN 47403 USA
1-800-288-4677   International: 001-812-330-2909

Because of the dynamic nature of the internet, any web addresses or links contained in this book may have changed since publication and may no longer be valid.

The views expressed in this work are solely those of the author and may not reflect the views of the publisher, and the publisher hereby disclaims any responsibility for them.

To the greatest man who has ever walked the surface of this earth, my father.

Copyright © 1994 & 2012 by Paul Gouda.

All rights reserved. No part of this book may be used, reproduced, translated, or transmitted in any form or by any means, electronic, graphical, or mechanical. This includes photocopying, recording, or using any information retrieval system, without a written permission from the author.

The preceding statement excludes the obviously permitted public use of the scientific facts utilized by this book, as well as the use of the scientific notes that have been published by the author via science sites and journals. The author also permits the use of short quotations of the material for reviews and non-commercial scholastic assignments.

ISBN: 978-1-4759-4802-8 (sc)
ISBN: 978-1-4759-4803-5 (ebk)

Printed in the United States of America.
Iuniverse rev. date: 08/30/2012

Canadian Intellectual Property Office.
Certificate of registration of copyright number: **1097823**
Paul Gouda, Owner & Author.
Issued pursuant to sections 49 & 53 of the Copyright Act.

Library and Archives Canada Cataloguing in Publication
Gouda, Paul
Arsenic, selenium, antimony : ultra trace analysis / Paul Gouda.
Includes index.
1. Arsenic--Analysis. 2. Selenium--Analysis. 3. Antimony--Analysis.
4. Ultratrace analysis. I. Title.

QP519.9.U48G68 2011 543 C2011-901291-X

American Library of Congress Control Number: 2012916152

Preface

The business of chemical analysis of environmental, medical and commercial samples has recently witnessed a large market demand as well as extensive scientific research and development.

Dr. Paul Gouda at his laboratory office at EPA during a media release report.

The days of high ppm level analysis done by a lab technician via a routine bench procedure involving mere volumetric titration, gravimetric precipitation and basic AAS at ppm level are long gone. Remember the days when we used to analyze, say, Hg, via such methods as diphenylthiocarbozone? Today's market demands qualified analytical chemists and state of the art instrumentation to deal with trace ppb and ppt detection level.

Lake water, seafood, soil, pharmaceutical products and many other samples are sent daily to analytical laboratories for several inorganic and organic analysis, and are subjected to improved regulations that specify the new acceptable limits.

Many may have heard about arsenic as a poisonous element we see in films, but many may not realize how common some elements such as arsenic, selenium and antimony are in nature, and in a variety of forms.

A few years ago, a lady friend was making coffee. As she ran out of filter papers, she innocently used a paper towel. The hot water went through the coffee and the acting filter paper as usual, but she never thought that paper towel could contain a high level of arsenic.

"Why would they add arsenic to paper towels?" she asked. "They don't." I explained. Products such as paper towels are made from recycled newspapers and other scrap wood sources and dry leaves and branches - which (like soil) may contain high levels of arsenic.

Products such as coffee paper-filters, are purified via such techniques as ion exchange and other chemical procedures - to extract arsenic. Paper towels, however, are meant for external use, and are not meant to be subjected to boiling water for one to then drink the extract! I explained.

It is of course true that the form of arsenic, selenium or antimony makes all the difference as to whether they are harmful to us or to what degree. It is also true that soaking a paper towel in boiling water may extract other chemicals that are not meant to be ingested; but that's a different subject. Our subject here is As, Se, Sb; and one must also understand that not every form of these elements is actually a threat to our body or can be absorbed into our body. Some of these forms, especially the metallic elements, are not particularly or practically harmful to humans, while other forms are easily absorbed into our body if ingested.

It is the harmful form of such parameter that is subjected here to our chemical study.

As I indicated, we are now more health conscious than ever, and the standard of specifications (though still needs to be tightened in North America to meet the better standard in Europe) - is still much higher than it used to be.

The authror at his Ph.D. graduation, UoT.

This paper deals with inorganic and organic parameters of arsenic, selenium and antimony.

These elements are common in nature: in soil, wood "trees" and wood-related products and in other sources; and in a variety of forms, both organic and inorganic.

Clearly agricultural soil is always a concern as vegetation will absorb such elements from the soil.

As well, potable water, residential sites, food and pharmaceutical products and others, each has its specific allowable limit of As, Se and Sb. Hence, the role of the analytical laboratory methodology and the appropriate analytical procedure to secure the suitable DL "detection level" or "detection limit" for the specific case, are a must..

The methods and the studies to follow were utilized, revolutionized and produced by Canadian chemist, Dr. Paul Gouda of Ontario, Canada; and have been well received and adopted by the scientific arena, including analytical laboratories.

The book also introduces five AAS "Atomic Absorption Spectroscopy" and ICP "Inductively Coupled Plasma" methods – with emphasis on Hydride Generation AAS.

These five methods were created by and have been named after the inventor chemist Dr. Paul Gouda.

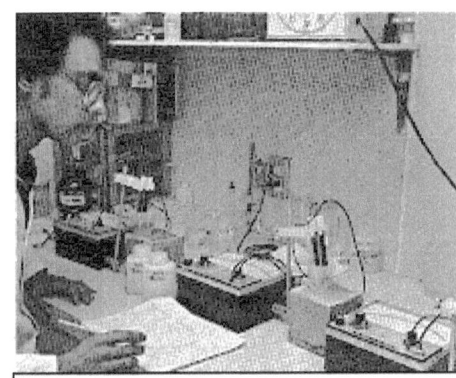

The author, at Barringer Research Magenta Int. Toronto, Ontario, reviewing a sample data reported by a chemist on a new ion release buffer for a blind electrode QC

This book was first produced in 1994 as a laboratory manual for bench chemists during the author's employment as a head chemist at B.R.M.

A large portion of the material was posted in 1998 – with the author's permission – by an online science society site.

It was then published in 2001 by "Great Canadian books" as a science reference.

This 3rd edition is published by Penguin Group, Iuniverse Bloomington in 2012.

..

Table of contents

Preface
SOP Introduction

Atomic Absorption -
Principal / theory
Hydride generation technique &
Graphite furnace "tube" techniques.
Instrumentation
Digestion scheme

Calibration & Correlation coefficiency
Reagents
Standards & special solutions
Quality control
Precautions

Analytical notes / synopsis
Oxidation states

Common compounds
Oxidation of selenides
Ferric selenite
Ferrous sulfate use
Selenium halides & selenosulfates
Selenides and selenites
Thiosulfate titration

Monohaliodes Se_2X_2, the dihalides SeX_2, the tetrahalides
Selenium oxyhalides $SeOX_2$
Permonoselenic acid H_2SeO_5, perdiselenic acid $H_2Se_2O_8$ and
Pyroselenic acid $H_2Se_2O_7$
Sodium selenocyanate NaSeCN Selenocyanogen SeCN / $(SeCN)_2$,

Selenium selenosulfate Na_2SeSO_3.
Arsenopyrite AsFeS

Tetrahedrally hybridized arsenic
Trigonal-bipyramidal or octahedral valence states

Permanganate method and the potassium bromate titration
Arsenious sulfide
Primarily orthoarsenic acid, H_3AsO_4
Arsine [hydrogen arsenide, arsenic trihydride]

Arsine hexahydrate $AsH_3.6H_2O$
Arsenic monohydride As_2H_2 or AsH
Tetrachlorophenylarsorane $C_6H_5AsCL_4$ and Dichlorophenylarsine $C_6H_5AsCL_2$.
Trihalides, but arsenic pentafluoride

Arsenic pentoxide As_2O_5
Hydrolysis of haloarsines
Hetrocyclic organoarsines
Arsines methylarsine CH_3AsH_2

Mimetite, $Pb5Cl(AsO3)3$
Erythrite, $CO3(AsO4)2·8H2O$
Proustite, $Ag3AsS3$
Specific halogen compounds with valence +3 and +5

Crystals of $H3AsO4(H2O)0.5$
Monosodium methyl arsenate
Arsenic pentoxide and tribromide
Arsenic trisulfide, trioxide and triiodide

   Arsenate minerals
   Arsenide minerals
   Stibine SbH_3
   Antimonate { SbO_4^{3-}} and sodium borohydride role
   Stibiconite $Sb_3O_6(OH)x$, cervantite Sb_2O_4 or Sb_2O_3 / Sb_2O_5
   Dicyclohexylstibine

Chlorodicyclohexylstibine
Phenylstibine and diphenylstibine
Diphenylstibine
The role of dialkylstibinic acids
Diazonium salt

Antimony pentachloride and aryltetrachloroantimony compound
Valentenite and senarmonite Sb_2O_3 and kermesite Sb_2S_3.
Trifluoromethylarsine CF_3AsH_2
Diarylantimony trichloride
Tetraantimony dichloride pentoxide

   Proustite
   Copper arsenate hydroxide or basic copper arsenate
   Monosodium methyl arsenate & Sodium arsenate
   Arsenic pentachloride
   Arsenic pentafluoride

   Arsenic pentoxide
   Arsenic tribromide
   Arsenic trifluoride
   Arsenic triiodide
   Arsenic trioxide

   Arsenic trisulfide
   Indium arsenide antimonide phosphide

   Arsenate minerals including:
   Aktashite
   Adamite
   Abernathyite
   Aerugite

   Agardite
   Akrochordite
   Alarsite
   Cabalzarite
   Andyrobertsite
   Annabergite

Arseniosiderite
Arsenoclasite
Arthurite
Austinite
Bayldonite
Beudandite

Brassiteis
Bukovskyite
Cahnite
Campylite
Chalcophyllite
Clinoclase
Conichalcite

Cornubite
Duftite
Euchroite
Fluckite
Fornacite
Geigerite
Haidingerite

Arthurite
Jarosewichite
Kaatialaite
Kankite
Lavendulan
Liroconite
Metazeunerite

Mimetite
Mixite
Olivenite
Pharmacosiderite
Picropharmacolite
Sewardite
Strashimirite
Scorodite

Strashimirite
Tsumcorite
Tyrolite
Vladimirite

Warikahnite
Weilite
Xanthiosite
Zeunerite
Arsenide minerals including:

algodonite Cu_6As
domeykite Cu_3As
löllingite $FeAs_2$
nickeline $NiAs$

rammelsbergite $NiAs_2$
safflorite $(Co,Fe)As_2$
skutterudite $(Co,Ni)As_3$
sperrylite $PtAs_2$

Selenium chalcogen compounds
Halogen compounds
Organoselenium compounds
Halides
Hetrocyclic organoarsines

Selenium trioxide
Selenium disulfide
Selenium dichloride
Tetraselenium tetranitride

Organoselenium
Diselenides
Diphenyldiselenide
Selenenyl halides
Perseleninic acids

Selenoxides
Selones
Vinylic selenides
Selenoxide oxidations

stibnite – trisulfide
stibiconite
cervantite
valentenite

senarmonite
kermesite
sesquioxide
valentenite

stibnite – trisulfide
stibiconite
cervantite
valentenite
senarmonite

kermesite
sesquioxide
valentenite
senarmonite
kermesite

stibiconite
Stibine
phenylstibine and diphenylstibine
hydrolysis of halo-and dihalostibines
dialkylstibinic acids

aryltetrachloroantimony compounds
Antimony trichloride
Trimethylstibine
Triphenylstibine
Antimony pentafluoride

Antimony potassium tartrate
Antimony sulfate,
Antimony telluride
Antimony tribromide
Antimony trichloride

Antimony triiodideis
Antimony triselenide
Isotopes of antimony
Halides

Organoantimony compounds
Oxides and hydroxides
Antimonide minerals
Antimonides:
Senarmonite
kermesite
Antimonides

Flame stoichiometry
Baseline noise level
Interference
Optimization

Ghraphite furnace Atomic absorption.
Hydride generation Atomic Absorption.

Methods developed by the author, adopted by EPA analytical laboratories, and globally named after him:

PGASSBHG   Paul Gouda As/Sb
            By hydride generation Atomic Absorption.

PGSEHG   Paul Gouda / Se / Hydride Generation

PGASSBHGA   Paul Gouda As/Sb/
             Graphite furnace Atomic Absorption

About the author

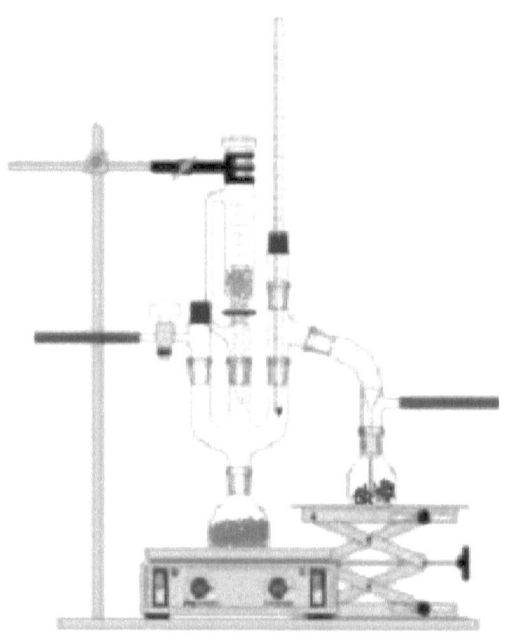

About this book.

The objective of this book is to provide the bench chemist with the following material:

- Information applicable to wet chemistry and precautions to assist with a safe sample treatment, whether titration, precipitation or general test-tube or beaker digestion.

- Information applicable to achieving accurate instrumental results, whether Atomic Absorption or Inductively Couples Plasma.

- Information helpful in the area of general safety precautions and expected reactions of by-products of secondary reactions.

- Information on factors and facts associated with the successful ultra-trace analysis.

The book will serve as a laboratory or bench analyst manual, analytical chemist reference and a university textbook.

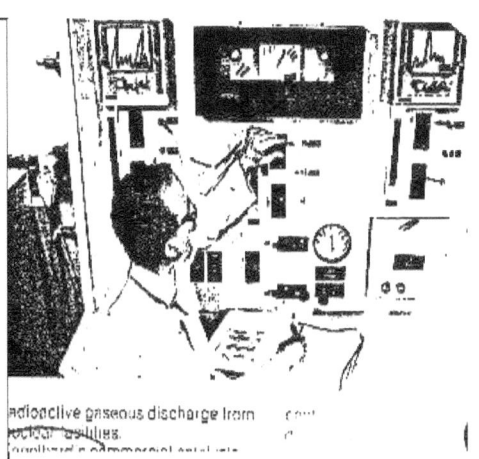

A report by the EPA journal featured an interview with Dr. Paul Gouda on ultra trace analysis of food samples and pharmaceutical drugs using a method developed by Dr. Gouda that was named, and is now known as {PGHgICP}. Dr. Gouda is seen here verifying a result utilizing a state of the art ICP.

The analytical study
"As / Se / Sb" SOPs

A complete *Standard Operating Procedure* addressing the laboratory chemical analysis of Arsenic, Selenium & Antimony in commercial, pharmaceutical and industrial samples.

Atomic Absorption - Hydride generation and Graphic furnace techniques.

This complete SOP reference paper covers the laboratory application, administration & technical operation aspects of the ultra trace analysis of As, Se, & Sb, and is presented as a specific illustrative study.

The methodology theory.

Atomic absorption notes.

Bench – wet chemistry applications

Quality control standards

Principal / theory

1.0

Metalloid elements {Se / As / Sb} are prepared in acidic medium to convert all forms of arsenic, selenium and antimony to arsenate {AsO_4^{3-}}, selenate {SeO_4^{2-}}, and antimonate {SbO_4^{3-}} react with sodium borohydride and are reduced to arsine, hydrogen selenide and stibine.

The volatile hydrides in the reaction tube are carried by an inert gas [argon] into a cell [quartz tube] that is heated by air / acetylene flame and is situated in the optical path of AAS where the gaseous hydrides are reduced to atomic species and are then determined by conventional atomic absorption.

The instrumental part involves the atomic absorption technique; hydride generation / graphite furnace AAS.

2.0- Instrumentation:

2.1 Description:

a} Hardware configuration:

The Atomic Absorption Spectrometers utilized in this study are:
Thermo-Jarrell Ash, Smith Heifta 22 AAS.
And: Varian VGA - 76 generation pump accessory.

b} Technical software:

The example used for data illustration is based on the software provided by the AAS manufacturer; in this case, PC Data system - TJA Thermospec S.W. by Thermo-Jarrel Ash.

3.0- Test codes:

3.1 Se-20-SO   {Sample type: Soil. Parameter : Selenium}

3.2 AsSb-20-SO {Sample type: Soil. Parameter : Arsenic & Antimony}

3.3 Preservation code: D.

3.4 Storage method : Refrigeration of non-aqueous samples and HNO_3 treatment of aqueous samples to pH < 2.

3.5 Container: 8

3.6 Target hold time: 1 month.

4.0- Method:

Se EPA-7741 / modification (Environment Protection Agency) As/Sb: EPA-7061 & 7042 / modification.

5.0- Applications:

5.1 Samples are digested in oxidizing acid mixture and are treated to:

> a) **reduce selenium in the digestate selenium IV as selenite $SeO_3^=$. Selenite is reduced to hydrogen selenide H_2Se by sodium borohydride "$NaBH_4$" and HCL. The hydrogen selenide is reduced in the flame [cell] to selenium atoms.**
>
> b) **convert all forms of arsenic to arsenate ion AsO_4^{3-} [The arsenate in the acidic solution is reduced by sodium borohydride to AsH_3 "arsine".] Atomic arsenic is then determined by conventional atomic absorption technique.**

5.2
 Hydride forming elements must be converted to the required oxidation state. e.g. Sb "V" and As "V" are reduced to Sb "III" and As "III' with KI {after HCL treatment}. While Sb reduction is spontaneous, As

reduction takes approximately 1 hour at room temperature or 5 minutes at 50 ^{0}C [water bath].

6.0- Digestion scheme:

6.1 A soil / sludge sample aliquot of 0.25 - 1.0 gm is treated with < 1 ml of H_2O and is left for 5 minutes in the 50 cc digestion tube. The sample is then treated with an acid mixture of 6:3:1 HNO_3 / H_2SO_4 / $HCLO_4$ and is allowed to sit for 10 minutes. 2-5 ml of acid mixture is the common practice depending on sample nature and need. 3 ml is a practical guideline. Solution samples are shaken and sampled by weight or volume as necessary.

6.2 The sample is digested on a heat block with a starting temperature of 70 ^{0}C and a maximum temperature of 250 ^{0}C. The temperature must be monitored and controlled based on sample reaction / nature / matrix to obtain optimum digestion. Depending on the digestion temperature and the sample nature, the digestion would take 4 - 16 hours on the hot plate [heat block]. A good guideline would be a 10 hours digestion at 120 ^{0}C. Special attention must be given to individual samples during digestion especially when the analyzing for Se or a parameter in elemental / volatile state.
All oxides of nitrogen / NOx must be expelled prior to instrumentation. An indication that traces of nitric are removed and the digestion is complete, is the reduction of the solution volume down to 1 cc and the production of floating - suspended white fumes of SO_3 are clearly observed.

6.3 The sample is removed off the heat block and let cool down to room temperature.
1 ml of H_2O is added followed with 5 ml of HCL and the sample is let sit for 10 minutes and is then heated in a water bath at 80 ^0C for 60 minutes.

6.4 The sample is removed from the water bath, let cool to room temperature and H_2O is added to make up volume to 50 cc. Shake well and divide the sample into 2 portions of 25 cc each by pouring 25 ml into a plastic centrifuge tube.

6.5 One set of the two 25 cc samples is for Se determination. The sample requires no further treatment and is ready for AAS determination with no further dilution; provided of course that the level monitored is within the optimum calibration graph coverage.

6.6 The other set of digested 25 cc sample is to be used for As / Sb determination {dual elements / cathode lamps - AAS setting }. Add 0.5 ml of 50 % KI w/v solution and heat in a water bath at 80 ^0C for 15 minutes.

6.7 Dilute samples 1:10 in graduated 12 cc plastic test tube as follows:

1 ml sample { after 6.6 treatment above } + 1 ml HCL + 0.2 ml KI solution 50% w/v and add 7.8 ml of H_2O to bring final volume to 10 cc. Shake well.

This portion of the sample is ready for AAS - As/Sb instrumentation. Prepare only when ready for AAS operation.

7.0- Calibration:

7.1 lower range operation:
Standard solutions / ppb: 0 (blank), 1, 3, 5 ug/l.

7.2 Higher range operation:
Standard solutions / ppb: 0 (blank), 1, 5, 10 ug/l.

7.3 Correlation coefficiency: 0.995 minimum. .998 is expected. .999 is sought.

8.0 Instrumental accessories (samples used in this case):

8.1: Varian VGA-76 peristaltic pump:

a] Rate {cc per min.} Hydrochloric acid: 1
Borohydride solution: 1
Sample: 8
b] Insert / purge gas: Ar {pressure: 50 psi}.
{Air supply: compressed - cylinder or filtered air supplied by a pump}
8.2 Burner: conventional air / acetylene
8.3 Nebulizer suction rate: H_2O: 5 cc min.

8.4 Cell: open ended quartz tube placed in the optical scope of AAS.
8.5 Other apparatus:

- Aluminum heating block with 40 holes

for test tubes.
- 50 cc digestion test tubes
- Temperature controlled hot plate
- Other necessary glassware

9.0 Reagents:

    9.1 H_2O: Distilled & Deionized

    9.2   HCL: neat - conc. analytical grade

    9.3 $NaBH_4$: Analytical grade

    9.4   NaOH: Analytical grade

    9.5 H_2SO_4 / HNO_3 / $HClO_4$:
        neat - conc. analytical grade

    9.6   K.I.: Analytical grade

10- Standards & special solutions:

10.1 Borohydride solution: to be pumped into the reaction tube:
0.9 % $NaBH_4$ + 0.8 % NaOH

{4.5 gm sodium borohydride + 4 gm sodium hydroxide to 500 ml with H_2O. The solution is filtered with #42 paper}.

Two more concentrations of borohydride solutions may be needed for the rare & special sample:
        a} 0.6% $NaBH_4$/0.5 NaOH
        b) 0.3% $NaBH_4$ & 0.2 N

The reason is explained in this literature. Dilution from the 0.9/0.8 solution is sufficient.

10.2   **Hydrochloric**:

a} to be pumped into the reaction tube:

50 ml H_2O + 300 ml HCL conc. {both 10 M & 5 M HCL may be needed as explained in this SOP}

b} conc./neat is used for other digestion as indicated.

10.3   **Blank sample**:

a] *Arbitrary zero*: 5 ml HCL - conc. + 45 ml H_2O {10% HCL v/v}.

b] *Analytical zero*: 1 ml H_2O + 3 ml of 6:3:1 nitric / sulfuric / perochloric acid mixture digested with the sample and treated in an identical manner {i.e. addition of 5 ml HCL and water bath treatment, dilution with H_2O, addition of KI for As / Sb determination ...etc.}

10.4   **Stock standard solution**:

100 mg/L [ppm] certified solution; e.g. Plazmachem.

10.5   **Working solution**:

100   ug/L [ppb] solution from certified standard prepared in 5 % HCL from 2 different sources, e.g. BDH.

10.6   **Calibration standards:**

- Blank - see above.
- 1 , 3 , 5 , 10 ug/L digested with and treated identically as the set of samples.
  0.5 , 1 , 1.5 , 2.5 & 5 ml of 100 ug/L std are pipetted into 50 cc digestion tubes when the set of samples are prepared for digestion.

10.7   **Standard addition:**

A spiking solution of 150 ug/L Se & 1500 ug/L As/Sb in 5% HCl. 1 ml is added to the spiked sample {Q3} prior to digestion so that it would undergo the same conditions & treatment of the samples. The recovery would then reflect on both digestion and matrix affect factors.

Since the final volume of the initial digestion is 50 cc, the concentration of the standard addition = 3 ppb Se & 30 ppb As/Sb. As/Sb digestion involves a further 1/10 dilution, resulting in a final spike value of +3 ppb in solution.

11- Quality control:

   11.1 **Digested Q.C.:**

a} Q1: A digested blank with identical treatment.
b} Q2: A digested 3 ug/l standard with identical treatment.
c} Q3: A spiked sample. A replicant of 1st sample "D" with an ideal concentration / recovery of sample "D" value + 3 ppb spike conc.

d} Q4: A replicant of "D" sample with no standard addition.

11.2 References:

a} Standard from a second / separate source for cross-check confirmation; e.g. BDH or Fisher Vs. Plazamachem or CGS.

b} Control samples; i.e. certified soil material - e.g. "Sewage sludge" each with a known concentration. The QC sample is digested & treated under conditions identical to the tested sample's.

c} In-house control std. from certified material such as As_2O_3, H_2SeO_3, $Fe(OH)SeO_3$, Sb_2O_3 and other certified salts or metals. {Checked against references "a" & "b" above for confirmation)

12- Precautions:

12.1 use a perochloric acid fume hood.

12.2 observe the safety practice of handling acids and harmful / toxic substances.

12.3 neutralize end-solutions with $CaCO_3$ or limestone chips prior to disposal.

Flame stoichiometry

Oxidation states

cationic and anionic related side-effects

Matrix complications

13- Analytical notes / synopsis:

The analyst must be familiar with certain facts and factors that play a major role in the successful hydride operation.

A discriminating - educated approach - as opposed to a blind routine application - makes the critical difference. The following factors require special attention:

- The matrix, e.g. soil / vegetation / alloy / fish / lake water... etc.

- The evaluated major compound forms, e.g. salt / inorganic metallic / organic volatile / complex compounds .. etc.

- Special - recognized presence [e.g. an oxide] easily determined by colour / simple bench tests / common or obvious reaction.

The analyst's adjusted approach to an individual sample includes these areas:

- Digestion treatment: temperature / duration / acid addition.

- Flame stoichiometry. The type of the sample and its major composition play a major role in controlling several chemical interference and atomization complications as well as several **cationic and anionic related side-effects** that results in serious spectral problems.
  A reasonable range of flame temperature is allowed to control how lean or rich the flame should be.
  Commonly, a temperature of $650\ ^0C$ is a good guideline. This is adjusted depending on the element

analyzed - whether it's Se, As or Sb - and based on the individual sample itself.

- AAS setting. This includes adjusting the spectral band width / holllow cathode current / gain voltage to best deal with a certain matrix. The objective is to deal with:

- baseline noise
- nonlinear curve
- reversal absorption
- non-absorbable radiation
- background problems
- basic chemical & spectral interference

The enclosed setting / AAS method is a good guideline. However, as with any AAS operation, a sample may need individual attention in order to be able to capture the right peak and to minimize spectral and chemical interference.

14- Traces of nitric acid remaining after digestion - especially after warming up the sample - is likely to result in considerable analytical interference especially with Se determination.

The digestion must therefore be completed until SO_3 fumes are observed. However, it's also essential that over-digestion be avoided otherwise the sample will be subject to losses of selenium {elemental or compound form} in a volatile state.

15- Oxidation states:

The addition of HCL & KI serves to reduce As/Sb/Se to valence states most favorable as follows:

Arsenic: preferred "As III" - alternate "As V". Reduction with KI. Heating needed.

Selenium: preferred as "Se IV" - alternate "Se VI". Reduction with 6-7 M HCL {or a higher conc.} with possible need for heating.

Antimony: preferred "Sb III" - alternate "Sb III". Reduction with KI - spontaneous.

Arsenic characteristics

The following selected fact will allow the bench chemist to recognize several compounds he'll run into during test-tube sample digestion, and will assist him to avoid acomplications, errors and accidents.

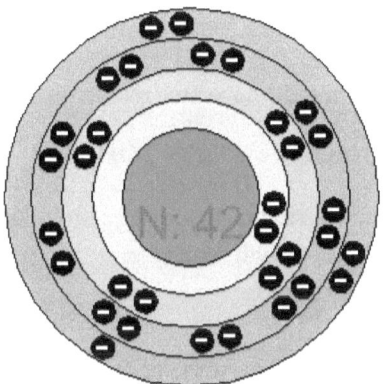

Arsenic tends to form the tetrahedral molecule As_4, in which, via sharing electrons, each As is bound to the other three covalently.

Phosphorus acts similarly. As_4 is a yellowish gas that condenses to a yellow, wax-like solid with a density around 2.0 g/cc. The gas has a distinctive garlic-like odor. This resembles white phosphorus. However, white – unlike yellow arsenic, phosphorus is reasonably stable. In the presence of light, yellow or γ-As rearranges itself into higher density and darker color, such as brown or black β-As of 4.7 density.

While there are several such structures, they all are nonmetallic, without lustre and no electrical conductivity. The most stable form at room temperature is metallic α-As, with grey color and metallic luster, which soon tarnishes on exposure to air. It has a 6.95 x 10-4 per °C coefficiency of expansion, and specific heat 0.0822 cal/g/°C. Though it is a metal, it is brittle and does not melt, but sublimes directly into As_4 at 615°C. It is analogous to red phosphorus - which is not a metal – but, arsenic can dress as a metal or a nonmetal, depending on how it aggregates, depending on the balance between lowest energy and highest entropy.

In the trioxide As_2O_3, As has valence +3, while in the pentoxide As_2O_5 the valence is +5. When these oxides are dissolved in water, they attract H^+ and OH^- ions and may rearrange their structures. Now, if we look at Na_2O, the oxide of the definitely clearly metallic sodium, adding water results in:

$Na_2O + 2H_2O \rightarrow 2NaOH \rightarrow 2Na^+ + 2OH^-$

The solution becomes basic because of the abundant OH^- ions. In the case of As_2O_5, we find that:

$As_2O_5 + 3H_2O \rightarrow 2H_3AsO_4$.

The molecule formed is orthoarsenic acid, which dissociates to give H^+ ions, and an acidic solution. This resembles orthophosphoric acid, H_3PO_4. Arsenic behaves as a nonmetal, like phosphorus, and can form salts with metals *"arsenate."*

Similarly, As_2O_3 hydrolyzes to H_3AsO_3, arsenious acid, whose salts are the *arsenites*. Isolating the arsenic acids in stable dry form (orthoarsenic acid seems to form a hydrated crystal) is not a feasible laboratory procedure. The oxides are reformed upon evaporation of the solutions. The arsenate ion, AsO_4^{---}, is very similar to the phosphate ion, PO_4^{---}. Arsenic is poisonous because the arsenate attempts unsuccessfully to replace phosphate in metabolic processes.

A few arsenate minerals are found, mostly in altered surface deposits. Mimetite, $Pb_5Cl(AsO_3)_3$, a soft, heavy mineral, is found in supergene-enriched zones. Scorodite, $FeAsO_4 \cdot 2H_2O$ (ferric arsenate) is soft and light, often in hot springs. Erythrite,

Co$_3$(AsO$_4$)$_2$·8H$_2$O (cobaltous arsenate) is a bright red mineral known as *cobalt bloom*. There is a series of such minerals in which Ni replaces Co, ending with *annabergite* or *nickel bloom*.

In most arsenic minerals, the arsenic replaces sulphur in more familiar minerals. The hard and heavy mineral NiAs, niccolite or kupfernickel, is an example. Silver white cobaltite, CoAsS, is hard and heavy and occurs in pyritohedrons, but is distinguished from pyrite by its colour. FeAsS, arsenopyrite, is perhaps the commonest arsenic mineral, and is also called mispickel. It is found in striated prisms. (Co,Ni,Fe)As$_3$, skutterudite, a mineral used to be called smaltite, contains varying quantites of the iron group metals.

Proustite, Ag$_3$AsS$_3$, or "ruby silver," is soft and of medium weight. Cu$_3$AsS$_4$ is black enargite, the source of arsenic in the mines of Butte, Montana. Note that in the last two compounds, S has replaced O in the arsenate and arsenite. Enargite is cuprous thioarsenate.

Arsenic acid is the chemical compound formula H3AsO4. Commonly known as AsO(OH)3. It is a colorless acid and it is the arsenic analogue of phosphoric acid. Arsenate and phosphate salts behave in a similar pattern. Arsenic acid is found in solution where it is largely ionized. Its hemihydrate form (H3AsO4·½H2O) forms stable crystals. Crystalline samples dehydrate with condensation at 100 °C.

Being a triprotic acid, its acidity is described by three equilibria:

H3AsO4 H2AsO + H+ (K1 = 10−2.19)
H2AsO HAsO + H+ (K2 = 10−6.94)
HAsO AsO + H+ (K3 = 10−11.5)

These Ka values are close to those for phosphoric acid. The highly basic arsenate ion is the product of the third ionization.

Unlike phosphoric acid, arsenic acid is oxidizing, demonstrated by its ability to convert iodide to iodine.

Arsenic acid is prepared by treating arsenic trioxide with concentrated nitric acid.

$$As_2O_3 + 2\ HNO_3 + 3\ H_2O \rightarrow 2\ H_3AsO_4(H_2O)0.5 + N_2O_3$$

The resulting solution is cooled to give colourless crystals of $H_3AsO_4(H_2O)0.5$. When the crystallisation is conducted at lower temperatures, the dihydrate crystallises.

General properties:

Name, symbol, number arsenic, As, 33
Element category   metalloid
Group, period, block 15, 4, p
Standard atomic weight 74.9216

[Ar] 4s2 3d10 4p3
2, 8, 18, 5

Electron configuration

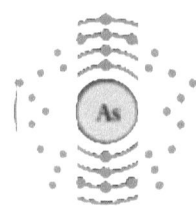

Physical properties:

Phase   solid
Density (near r.t.)   5.727 g·cm−3
Liquid density at m.p.   5.22 g·cm−3
Sublimation point   1137 °F   615 °C,   887 K,
Triple point   1090 K (817°C), 3628 kPa

Critical point	1673 K, ... MPa
Heat of fusion	(grey) 24.44 kJ·mol−1
Heat of vaporization	34.76 kJ·mol−1
Molar heat capacity	24.64 J·mol−1·K−1

Atomic properties:

Oxidation states	5, 3, 2, 1, -3 (mildly acidic oxide)
Electronegativity	2.18 (Pauling scale)
Ionization energies (more)	1st: 947.0 kJ·mol−1
	2nd: 1798 kJ·mol−1
	3rd: 2735 kJ·mol−1
Atomic radius	119 pm
Covalent radius	119±4 pm
Van der Waals radius	185 pm
Magnetic ordering	diamagnetic
Electrical resistivity	(20 °C) 333 nΩ·m
Thermal conductivity	50.2 W·m−1·K−1
Young's modulus	8 GPa
Bulk modulus	22 GPa
Mohs hardness	3.5
Brinell hardness	1440 MPa

Isotopes of arsenic

iso	NA	half-life	DM	DE (MeV)	DP
73As	syn	80.3 d	ε	-	73Ge
			γ	0.05D, 0.01D, e	-
74As	syn	17.78 d	ε	-	74Ge
			β+	0.941	74Ge
			γ	0.595, 0.634	-
			β−	1.35, 0.717	74Se
75As	100%	75As is stable with 42 neutrons			

Arsenic forms a number of halogen compounds with valence +3 and +5. For example, arsenious chloride $AsCl_3$, an oily liquid that decomposes when added to water; and AsF_5, arsenic fluoride, a colourless gas. The "ic" denoting the higher valence state conflicts somewhat with the attributive "ic" in the name. There is also As_2S_5, arsenic (arsenicic) sulphide. These compounds are not found in nature.

The gas AsH_3 or arsine is analogous to NH_3, ammmonia, and PH_3, phosphine. This compound is highly toxic, which was brought out forcefully soon after its discovery by the death of the prominent chemist Gehlen of Munich during experiments with it. In arsine, the conventional valence of arsenic is -3. If arsine is produced somehow, strong heating will decompose it into arsenic and hydrogen, and the arsenic vapor will condense as a shiny film of metallic arsenic, the "arsenic mirror," at a colder place.

This behaviour is used in the Marsh test for arsenic. The material to be tested is put in a flask with granulated zinc and hydrochloric acid is added. Any arsenic is changed to arsine as hydrogen is evolved and passes off in the evolved gas. The tube carrying the hydrogen is strongly heated at a certain point. If arsenic is present, an arsenic mirror is formed on the cooler part of the tube farther on. The hydrogen can also be lighted at the end of the tube. A cold surface held above the flame will also condense any arsenic that is formed. Hydrogen flames contain no carbon, and are quite invisible, even in a dark room. Unless you are very careful, these flames can be dangerous. Arsine generally oxidizes readily in the atmosphere.

A closed flask containing some arsenic can divert the experimenter. If the arsenic deposit is heated, it sublimes and condenses at some cooler part of the flask. The arsenic can be chased around the flask in this way with a flame.

***

Arsenic interferes with cellular longevity by allosteric inhibition of an essential metabolic enzyme pyruvate dehydrogenase(PDH) complex, which catalyzes the oxidation of pyruvate to acetyl-CoA by NAD+. With the enzyme inhibited, the energy system of the cell is disrupted resulting in a cellular apoptosisepisode.

Biochemically, arsenic prevents use of thiamine resulting in a clinical picture resembling thiamine deficiency. Poisoning with arsenic can raise lactate levels and lead to lactic acidosis. Low potassium levels in the cells increases the risk of experiencing a life-threatening heart rhythm problem from arsenic trioxide. Arsenic in cells clearly stimulates the production of hydrogen peroxide (H_2O_2).

When the H_2O_2 reacts with certain metals such as iron or manganese it produces a highly reactive hydroxyl radical. Inorganic arsenic trioxidefound in ground water particularly affects voltage-gated potassium channels, disrupting cellular electrolytic function resulting in neurological disturbances, cardiovascular episodes such as prolonged QT interval, neutropenia, high blood pressure, central nervous system dysfunction, anemia, and death. Arsenic is a ubiquitous element present in American drinking water.

Arsenic exposure plays a key role in the pathogenesis of vascular endothelial dysfunction as it inactivates endothelial nitric oxide synthase, leading to reduction in the generation and bioavailability of nitric oxide.

In addition, the chronic arsenic exposure induces high oxidative stress, which may affect the structure and function of cardiovascular system. Further, the arsenic exposure has been noted to induce atherosclerosis by increasing the platelet aggregation and reducing fibrinolysis. Moreover, arsenic exposure may cause arrhythmia by increasing the QT interval and accelerating the cellular calcium overload. The chronic exposure to arsenic upregulates the expression of tumor necrosis factor-α, interleukin-1, vascular cell adhesion molecule and vascular endothelial growth factor to induce cardiovascular pathogenesis.

***

Tissue culture studies have shown that arsenic blocks both IKr and Iks channels and, at the same time, activates IK-ATP channels. Arsenic also disrupts ATP production through several mechanisms. At the level of the citric acid cycle, arsenic inhibits pyruvate dehydrogenaseand by competing with phosphate it uncouples oxidative phosphorylation, thus inhibiting energy-linked reduction of NAD+, mitochondrial respiration, and ATP synthesis.

Hydrogen peroxide production is also increased, which might form reactive oxygen species and oxidative stress. These metabolic interferences lead to death from multi-system organ failure, probably from necrotic cell death, not apoptosis. A post mortem reveals brick red colored mucosa, due to severe hemorrhage. Although arsenic causes toxicity, it can also play a protective role.

Arsenate can replace inorganic phosphate in the step of glycolysis that produces 1,3-bisphosphoglycerate from glyceraldehyde 3-phosphate. This yields 1-arseno-3-phosphoglycerate instead, which is unstable and quickly hydrolyzes, forming the next intermediate in the pathway, 3-phosphoglycerate. Therefore glycolysis proceeds, but the ATP molecule that would be generated from 1,3-bisphosphoglycerate is lost - arsenate is an uncoupler of glycolysis, explaining its toxicity.

As with other arsenic compounds, arsenate can also inhibit the conversion of pyruvate into acetyl-CoA, blocking the Krebs cycle and therefore resulting in further loss of ATP.

Some species of bacteriaobtain their energy by oxidizing various fuels while reducing arsenates to form arsenites. The enzymes involved are known as arsenate reductases.

Just like ordinary photosynthesis uses water as electron donor, producing molecular oxygen, bacteria employs photosynthesis with arsenites as electron donors, producing arsenates . Research shows that photosynthesizing organisms produced the arsenates that allowed the arsenate-reducing bacteria to thrive.

Another research cultured samples of arsenic-resistant GFAJ-1 bacteria from a lake, using a medium high in arsenate and low in phosphate concentration. The findings suggest that the bacteria may partially incorporate arsenate in place of phosphate in some biomolecules, including DNA.

***

Being a triprotic acid, its acidity is described by three equilibria:

$$H_3AsO_4 \rightleftharpoons H_2AsO_4^- + H^+ \quad (K_1 = 10^{-2.19})$$
$$H_2AsO_4^- \rightleftharpoons HAsO_4^{2-} + H^+ \quad (K_2 = 10^{-6.94})$$
$$HAsO_4^{2-} \rightleftharpoons AsO_4^{3-} + H^+ \quad (K_3 = 10^{-11.5})$$

These K_a values are close to those for phosphoric acid. The highly basic arsenate ion is the product of the third ionization. Unlike phosphoric acid, arsenic acid is oxidizing, illustrated by its ability to convert iodide to iodine.

Arsenic acid is prepared by treating arsenic trioxide with concentrated nitric acid.

$$As_2O_3 + 2\ HNO_3 + 3\ H_2O \rightarrow 2\ H_3AsO_4(H_2O)_{0.5} + N_2O_3$$

The resulting solution is cooled to give colourless crystals of $H_3AsO_4(H_2O)_{0.5}$. When the crystallisation is conducted at lower temperatures, the dihydrate crystallises.

It also forms upon dissolving arsenic pentoxide with water, which is slow. It is also formed when meta - or pyroarsenic acid is treated with cold water. Arsenic acid can be prepared by reacting moist elemental arsenic with ozone.

$$2\ As + 3\ H_2O + 5\ O_3 \rightarrow 2\ H_3AsO_4 + 5\ O_2$$

The synthesis of calcium arsenate is commonly prepared by from disodium hydrogen arsenate and calcium chloride:

$$2\ Na_2H[AsO_4] + 3\ CaCl_2 \rightarrow 4\ NaCl + Ca_3[AsO_4]_2 + 2\ HCl$$

The solubility of this compound was found by Tartar, et al. to be 0.014g in 100g of water at 25°C.

Nearly a century ago it was made in large vats by mixing 4 moles of Calcium Oxide with 1 mole of Arsenic Oxide.

***

Chromated copper arsenate (referred to as CCA) is a wood preservative. It is a mix of chromium, copper and arsenic (as Copper(II) arsenate) formulated as oxides or salts. It preserves the wood from decay fungi, wood attacking insects, including termites, and marine borers. It also improves the weather-resistance of treated timber and may assist paint adherence in the long term.

CCA is widely used around the world as a heavy duty preservative, often as an alternative to creosote, and pentachlorophenol. Other water-borne preservatives like CCA include alkaline copper quaternary compounds (ACQ), copper azole (CuAz), ammoniacal copper zinc arsenate (ACZA), copper citrate, and copper HDO (CuHDO)

Recognized for the greenish tint it imparts to timber, CCA is a preservative that has been extremely common for many decades. Over time small amounts of the CCA chemicals, mainly the arsenic, may leach out of the treated timber.

***

Copper arsenate hydroxide or **basic copper arsenate** $\{Cu(OH)AsO_4\}$ is found naturally as the mineral olivenite. It is used as an insecticide, fungicide, and miticide.

Lead hydrogen arsenate, also called **lead arsenate**, **acid lead arsenate** or **LA**, chemical formula $PbHAsO_4$, is an inorganic insecticide used primarily against the potato beetle.

It is usually produced using the following reaction:

Pb(NO₃)₂(aq) +H₃AsO₄(aq) → PbHAsO₄(s) +2HNO₃(aq)

Lead arsenate was the most extensively used arsenical insecticide. Two principal formulations of lead arsenate were marketed: basic lead arsenate ($Pb_5OH(AsO_4)_3$), and acid lead arsenate $PbHAsO_4$.

Monosodium methyl arsenate(MSMA) is an arsenic-based herbicide and fungicide. It is an organic arsenate; but it is a less toxic organic form of arsenic, which has replaced the role of lead hydrogen arsenate in agriculture. It is one of the most common herbicides used on golf courses. It is typically used for control of grassy weeds such as crabgrass. Another common use is for burning the marked lines into grassy sports fields.

Sodium arsenate is the inorganic compound with the formula $NaH_2AsO_4·H_2O$. Related salts are also called sodium arsenate, including Na_2HAsO_4 and Na_3AsO_4. This particular salt is a colorless solid that is highly toxic. The potassium salt has similar properties.

The salt is the conjugate base of arsenic acid:

$$H_3AsO_4 \rightleftharpoons H_2AsO_4^- + H^+ \quad (K_1 = 10^{-2.19})$$

It is prepared in this way and crystallized by cooling a hot saturated aqueous solution, where it is highly soluble when hot (75.3 g in 100 mL at 100 °C). A fine white powder of the monohydrate crystallizes. Upon heating, the solid loses water of crystallization and then the dihydrogenarsenate anions condense to give $Na_2H_2As_2O_7$ and, above 230 °C, $NaAsO_3$.

Arsenic pentachloride is a chemical compound of arsenic and chlorine. This compound was first prepared in 1976 through the UV irradiation of arsenic trichloride, $AsCl_3$, in liquid chlorine at −105°C.[2] $AsCl_5$ decomposes at around −50°C. The structure of the solid was finally determined in 2001. $AsCl_5$ is similar to phosphorus pentachloride, PCl_5 in having a trigonal bipyramidalstructure where the equatorial bonds are shorter than the axial bonds (As-Cl$_{eq}$ = 210.6pm, 211.9 pm; As-Cl$_{ax}$= 220.7pm).

The pentachlorides of the elements above and below arsenic in group 15, phosphorus pentachloride and antimony pentachlorideare much more stable and the instability of AsCl₅ appears anomalous. The cause is believed to be due to incomplete shielding of the nucleus in the elements following the first transition series (i.e. gallium, germanium, arsenic, selenium and bromine) which leads to stabilisation of their 4s electrons making them less available for bonding. This effect has been termed d-block contractionand is similar to the f-block contraction normally termed the lanthanide contraction.

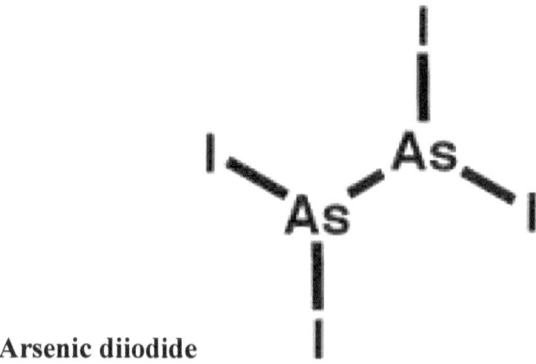

Arsenic diiodide

Arsenic pentafluoride can be prepared by direct combination of arsenic and fluorine

$$2As + 5F_2 \rightarrow 2AsF_5$$

It can also be prepared by the reaction of arsenic trifluoride and fluorine:

$$AsF_3 + F_2 \rightarrow AsF_5$$

Arsenic pentafluoride is a colourless gas and has a trigonal bipyramidal structure. In the solid state the axial As-F bond lengths are 171.9 pm and the equatorial 166.8 pm.

Arsenic pentafluoride forms halide complexes and is a powerful acceptor as shown by the reaction with sulfur tetrafluoride forming an ionic complex.

$$AsF_5 + SF_4 \rightarrow SF_3^+ + AsF_6^-$$

Arsenic pentoxide is the inorganic compound with the formula As_2O_5. This glassy, white solid is relatively unstable, consistent with the rarity of the As(V) oxidation state. More common, and far more important commercially, is arsenic(III) oxide (As_2O_3).

Arsenic pentoxide can be crystallized by heating As_2O_3 under oxygen. In fact, arsenic pentoxide decomposes to oxygen and As_2O_3 upon heating.

As_2O_5, which like phosphorus pentoxide is hygroscopic, dissolves readily in water to form arsenic acid, H_3AsO_4.

Arsenic tribromide is the inorganic compoundwith the formula $AsBr_3$. This pyramidal molecule is the only known binary arsenic bromide. $AsBr_3$ is noteworthy for its very high refractive index of approximately 2.3. It also has a very high diamagnetic susceptibility. The compound exists as colourless deliquescent crystals that fume in moist air.

Preparation:

Arsenic tribromide can be prepared by the direct bromination of arsenic powder. Alternatively arsenic(III) oxide can be used as the precursor in the presence of elemental sulfur:

$$2\ As_2O_3 + 3\ S + 6\ Br_2 \rightarrow 4\ AsBr_3 + 3\ SO_2$$

While $AsBr_5$ has not been well studied, the corresponding phosphorus compound PBr_5 is well characterized. $AsBr_3$ is the parent for a series of hypervalent anionic bromoarsenates including $[As_2Br_8]^{2-}$, $[As_2Br_9]^{3-}$, and $[As_3Br_{12}]^{3-}$.

Organoarsenic bromides, $(CH_3)_2AsBr$ and $(CH_3)AsBr_2$ are formed efficiently by the copper-catalyzed reaction of methyl bromidewith hot arsenic metal. This synthesis is similar to the direct process used for the synthesis of methyl chlorosilanes.

***

Arsenic trichloride is an inorganic compound with the formula $AsCl_3$, also known as arsenous chloride or butter of arsenic. This poisonous oil is colourless, although impure samples may appear yellow. It is an intermediate in the manufacture of organoarsenic compounds

$AsCl_3$ is a pyramidal molecule with C_{3v} symmetry. The As-Cl bond is 2.161 Å and the angle Cl-As-Cl is 98°25'±30.[2][3] $AsCl_3$ has four normal modes of vibration: $v1(A_1)$ 416, $v2(A_1)$ 192, $v3$ 393, and $v4(E)$ 152 cm^{-1}. Arsenic trichloride contains predominantly covalent bonds, which explains its low melting point.

This colourless liquid is prepared by treatment of arsenic(III) oxide with hydrogen chloridefollowed by distillation:

$$As_2O_3 + 6\ HCl \rightarrow 2\ AsCl_3 + 3\ H_2O$$

It can also be prepared by chlorination of arsenic at 80-85 °C, but this method requires elemental arsenic.

$$2\ As + 3\ Cl_2 \rightarrow 2\ AsCl_3$$

Arsenic trichloride can also be prepared by the reaction of arsenic oxide and sulfur monochloride. This method requires simple apparatus and proceeds efficiently:

$$2\ As_2O_3 + 6\ S_2Cl_2 \rightarrow 4\ AsCl_3 + 3\ SO_2 + 9\ S$$

Hydrolysis with water gives arsenous acid and hydrochloric acid:

$$AsCl_3 + 3\ H_2O \rightarrow As(OH)_3 + 3\ HCl$$

Although $AsCl_3$ is less moisture sensitive than PCl_3, it still fumes in moist air.

$AsCl_3$ undergoes redistribution upon treatment with As_2O_3 to give the inorganic polymer AsOCl. With chloride sources, $AsCl_3$, forms salts containing the anion $[AsCl_4]^-$. Reaction with potassium bromide and potassium iodide give arsenic tribromide and arsenic triiodide, respectively.

$AsCl_3$ is useful in organoarsenic chemistry, for example triphenylarsine is derived from $AsCl_3$:

$$AsCl_3 + 6\ Na + 3\ C_6H_5Cl \rightarrow As(C_6H_5)_3 + 6\ NaCl$$

Arsenic trifluoride is a chemical compound of arsenic and fluorine with the formula AsF_3. It is a colorless liquid which reacts readily with water.

Preparation and properties:

It can be prepared by reacting hydrogen fluoride, HF, with arsenic trioxide:

$$6HF + As_2O_3 \rightarrow 2AsF_3 + 3H_2O$$

It has a pyramidal molecular structure in the gas phase which is also present in the solid. In the gas phase the As-F bond length is 170.6 pm and the F-As-F bond angle 96.2°.

Arsenic trifluoride is used as fluorinating non-metal chlorides to fluorides, in this respect it is less reactive than SbF_3.
Salts containing AsF_4^- anion can be prepared for example $CsAsF_4$. the potassium salt KAs_2F_7 prepared from KF and AsF_3 contains AsF_4^- and AsF_3 molecules with evidence of interaction between the AsF_3 molecule and the anion.
With SbF_5 the ionic adduct $AsF_2^+\ SbF_6^-$ is produced.

Arsenic triiodide is the inorganic compoundwith the formula AsI_3. It is a dark red solid that readily sublimes. It is a pyramidal molecule that is useful for preparing organoarsenic compounds.

Preparation:

It is prepared by a reaction of arsenic trichloride and potassium iodide:

$$AsCl_3 + 3KI \rightarrow AsI_3 + 3\ KCl$$

Hydrolysis occurs only slowly in water forming arsenic trioxide and hydroiodic acid. The reaction proceeds via formation of arsenous acid which exists in equilibrium with hydroiodic acid.

The aqueous solution is highly acidic, pH of 0.1N solution is 1.1. It decomposes to arsenic trioxide, elemental arsenic and iodine when heated in air at 200 °C. The decomposition, however, commences at 100 °C and occurs with the liberation of iodine.

***

Arsenic trioxide is the inorganic compound with the formula As_2O_3. This commercially important oxide of arsenic is the main precursor to other arsenic compounds, including organoarsenic compounds. Approximately 50,000 tonnes are produced annually. Many applications are controversial given the high toxicity of arsenic compounds.

Preparation and properties:

Arsenic trioxide can be generated via many routine processing of arsenic compounds including the oxidation (combustion) of arsenic and arsenic-containing minerals in air. Illustrative is the roasting of orpiment, a typical arsenic sulfide ore.

$$2\,As_2S_3 + 9\,O_2 \rightarrow 2\,As_2O_3 + 6\,SO_2$$

Most arsenic oxide is, however, obtained as a volatile by-product of the processing of other ores. For example, arsenopyrite, a common impurity in gold- and copper-containing ores, liberates arsenic trioxide upon heating in air. The processing of such minerals has led to numerous cases of poisonings.

Arsenic trioxide is an amphoteric oxide, and its aqueous solutions are weakly acidic. Thus, it dissolves readily in alkaline solutions to give arsenites. It is less soluble in acids, although it will dissolve in hydrochloric acid, giving chloro compounds, ultimately arsenic trichloride with concentrated acid. Only with strong oxidizing agents such as ozone, hydrogen peroxide, and nitric acid does it give arsenic pentoxide, As_2O_5. Reduction gives elemental arsenic or arsine (AsH_3) depending on conditions. In this regard, arsenic trioxide differs from phosphorus trioxide, which readily combusts to phosphorus pentoxide.

Structure:

In the liquid and in the gas phase below 800 °C, arsenic trioxide has the formula As_4O_6 and is isostructural with P_4O_6. Above 800 °C As_4O_6 significantly dissociated into molecular As_2O_3, which adopts the same structure as N_2O_3. Three forms (polymorphs) are known in the solid state: cubic As_4O_6, containing molecular As_4O_6, and two related polymeric forms. The polymers, which both crystallized as monoclinic crystals, feature sheets of pyramidal AsO_3 units that share O atoms.

Arsenic trisulfide is the inorganic compound with the formula As_2S_3. This bright yellow solid is a well known mineral orpiment (Latin: auripigment), has been used as a pigment, and has played a role in the analysis of arsenic compounds.

This chalcogenidematerial is a group V/VI, intrinsic p-type semiconductor and exhibits photo-induced phase-change properties. The other principal arsenic sulfide is realgar, As_4S_4, which is red-orange and also occurs as a mineral.

As_2S_3 occurs both in crystalline and amorphous forms. Both forms feature polymeric structures consisting trigonal pyramidal As(III) centres linked by sulfide centres. The sulfide centres are two-fold coordinated to two arsenic atoms. In the crystalline form, the compound adopts a ruffled sheet structure.

The bonding between the sheets consists of van der Waals forces. The crystalline form is usually found in geological samples. Amorphous As_2S_3 does not possess a layered structure but is more highly cross-linked. Like other glasses, there is no medium or long-range order, but the first co-ordination sphere is well defined. As_2S_3 is a good glass former and exhibits a wide glass-forming region in its phase diagram.

Synthesis and reactions:

Amorphous As_2S_3 is obtained via the fusion of the elements at 390 °C. Rapid cooling of the reaction melt ensures the disordered arrangement of the bonds, resulting in the glass. The reaction can be represented with the chemical equation:

$$2 \text{ As} + 3 \text{ S} \rightarrow \text{As}_2\text{S}_3$$

Upon heating in a vacuum, polymeric As_2S_3 "cracks" to give a mixture of molecular species, including molecular As_4S_6. As_4S_6 adopts the adamantane geometry, like that observed for P_4O_6 and As_4O_6. When a film of this material is exposed to an external energy source such as thermal energy (via thermal annealing), electromagnetic radiation (i.e. UV lamps, lasers, electron beams), As_4S_6 polymerizes:

$$2/n \ (As_2S_3)_n \quad As_4S_6$$

Aqueous precipitation:

As_2S_3 forms when aqueous solutions containing As(III) are treated with H_2S. Arsenic was in the past analyzed and assayed by this reaction, which results in the precipitation of As_2S_3, which is then weighed. As_2S_3 can even be precipitated in 6M HCl. As_2S_3 is so insoluble that it is not toxic.

As_2S_3 characteristically dissolves upon treatment with aqueous solutions containing sulfideions.

$$As_2S_3 + 6 \text{ NaSH} \rightarrow 2 \text{ AsS}3-3 + 3 \text{ H}_2\text{S}$$

As_2S_3 is the anhydride of the hypothetical thioarsenous acid, $As(SH)_3$. Upon treatment with polysulfide ions, As_2S_3 dissolves to give a variety of species containing both S-S and As-S bonds. One derivative is S_7As-S^-, a ring that contains an exocyclic sulfido center attached to the As atom. As_2S_3 also dissolves in strongly alkaline solutions to give a mixture of AsS3–3 and AsO3-3

Reactions with oxygen:

"Roasting" As_2S_3 in air gives volatile, toxic derivatives, this conversion being one of the hazards associated with the refining of heavy metal ores:

$$2\ As_2S_3 + 9\ O_2 \rightarrow As_4O_6 + 6\ SO_2$$

***

Due to its high refractive index of 2.45 and its large Knoop hardness compared to organic photoresists, As_2S_3 has been investigated for the fabrication of photonic crystals with a full-photonic band-gap. Advances in laser patterning techniques such as three-dimensional direct laser writing (3-D DLW) and chemical wet-etching chemistry, has allowed this material to be used as a photoresist to fabricate 3-D nanostructures.

As_2S_3 has been investigated for use as a high resolution photoreist material since the early 1970s, using aqueous etchants. Although these aqueous etchants allowed for low-aspect ratio 2-D structures to be fabricated, they do not allow for the etching of high aspect ratio structures with 3-D periodicity. Certain organic reagents, used in organic solvents, permit the high-etch selectivity required to produce high-aspect ratio structures with 3-D periodicity.

Medical applications:

As_2S_3 and As_4S_4 have been investigated as treatments for acute promyelocytic leukemia (APL). The mode of action is thought to be similar to that for As_2O_3.

Ancient Egyptians reportedly used orpiment, natural or synthetic, as a pigment in artistry and cosmetics.

***

Precipitation of arsenic trisulfide is used as an analytical test for presence of dissimilatory arsenic-reducing bacteria (DARB).

As_2S_3 is so insoluble that its toxicity is low. Aged samples can contain substantial amounts of arsenic oxides, which are soluble and therefore highly toxic.

***

Indium arsenide antimonide phosphide (InAsSbP) is a semiconductor material.

InAsSbP has been widely used as blocking layers for semiconductor laser structures, as well as for the mid-infrared light-emitting diodes, photodetectors and thermophotovoltaic cells.

InAsSbP layers can be grown by heteroepitaxy on indium arsenide, gallium antimonide and other materials. The vibrational properties of the alloy has been investigated by Raman spectroscopy.

Arsenic

notes

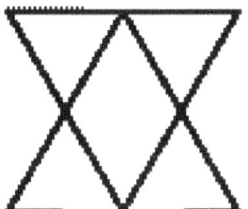

- Arsenic is common in soil samples in such mineral format as arsenopyrite AsFeS. Arsenic may be detected as a yellow sulfide As_2S_3 by precipitation from HCL solution.

The most common oxidation states of arsenic are -3, +3 and +5. However, recorded data indicates that in the majority of arsenic compounds the arsenic atom is in the tetrahedral valence state. Compounds in which the arsenic atom is three coordinate are assumed to contain the tetrahedrally hybridized arsenic atom with a lone pair of electrons in one of the hybird orbitals(1). Since the valence shell of the arsenic atom contains "d" orbital, compounds are known in which the arsenic atom adopts trigonal-bipyramidal or octahedral valence states.

- Metallic As is introduced in soil samples by carbon reduction of arsenious oxide As_4O_6 or by thermal decomposition to arsine and FeS of naturally occurring arsenical pyrite FeAsS. Arsine AsH_3 is not very stable to air oxidation.

In the +3 state arsenic forms arsenious oxide As_4O_6 producing a slightly acid solution which is believed to contain hydroxide $As(OH)_3$ or H_3AsO_3, [$HAsO_2$]. The hydroxide is amphoteric; it neutralizes acids to give solutions containing $As(OH)_2^+$, and it neutralizes bases to give solutions containing arsenite ions [$H_2AsO_3^-$, AsO_2^-, or $As(OH)_4^-$]:

$$As(OH)_3 + H_3O^+ \rightarrow As(OH)_2+ + H_2O$$

$$As(OH)_3 + OH^- \rightarrow H_2AsO_3^- + H_2O$$

- Any reaction in the sample that would produce H_2S into the arsenious solution, results in a yellow precipitate forming [commonly suspected to be solely As_2S_3 but experiments indicate that it probably has the molecular formula As_4S_6.]

Arsenious sulfide has great tendency to form colloids stabilized by adsorption of negative ions. These colloids can be coagulated by addition of H_3O^+ or other positive ions. If the H_2S was introduced into an arsenic acid [instead of arsenious solution] the yellow precipitation still appears, however in this case it is believed to be As_2S_5 [many chemists believe it to be As_4S_{10}].

- In the +5 state, the principal compounds of arsenic are arsenic acid and its derivatives, the arsenates. Arsenic acid is primarily orthoarsenic acid, H_3AsO_4

- The common arsenic hydrides are As_2H_4, As_2H_2 [or AsH] and As_4H_2, however, the only well-characterized binary compound of arsenic and hydrogen is arsine.

Arsine [hydrogen arsenide, arsenic trihydride], AsH_3 is a highly toxic colorless gas with an unpleasant garlic-like odor.

- Arsine is soluble in water and accepts a proton from water to form an onium ion as does ammonia. At temperature below -10 ^0C or under pressure arsine hexahydrate $AsH_3.6H_2O$ is formed.

- Chlorine reacts with arsine to give hydrogen chloride and arsenic. However, at low temperature the action of chlorine upon arsine produces chloroarsines AsH_2CL and $AsHCL_2$ {unstable yellow solids.}

The reduction of arsenic (III) compounds by stannous chloride in hydrochloric acid yields a brown amorphous powder corresponding to the formula of arsenic monohydride As_2H_2 or AsH.

Arsenic halides

Arsenic forms a complete series of trihalides, but arsenic pentafluoride is the only simple pentahalide known. Chlorine reacts with very cold arsenic trifluoride to produce a hygroscopic solid compound, arsenic dichloride trifluride $AsCL_2F_3$ consisting of $AsCL_4^+$ and AsF_6^- ions. Arsenic trichloride [arsenic III chloride] $AsCL_3$, the common halide of arsenic, is formed by spontaneous combination of the elements and, in addition, by reactions of chlorine with arsenic trioxide; or sulfur monochloride S_2CL_2 {or a mixture of S_2CL_2 and chlorine} with arsenic trioxide; or arsenic trioxide with concentrated hydrochloric acid.

Arsenic diiodide, a red solid, with water causes a disproportionation with the formation of arsenic and arsenic trioxide:

$$3 As_2I_4 \rightarrow 4 AsI_3 + 2 As$$

- Arsenic oxides and acids:

Arsenic pentoxide As_2O_5 is thermally unstable. Commercial arsenic acid corresponds to the composition, one mole arsenic pentoxide to four moles of water {arsenic acid hemihydrate $H_3AsO_4.0.5H_2O$}.

Arsenates are oxidizing agents and are reduced by conc. HCL.

Hydrolysis of haloarsines gives arsinous acids, R_2AsOH, or their anhydrides $(R_2As)_2O$.

Hetrocyclic organoarsines

Organoarsenic compounds are derived from arsine by replacing one, two or three hydrogens by an alkyl, cycloalkyl, aryl or even heterrocyclic group b.

Examples are Tetrachlorophenylarsorane $C_6H_5AsCL_4$ and dichlorophenylarsine $C_6H_5AsCL_2$.

The two primary arsines methylarsine CH_3AsH_2 and trifluoromethylarsine CF_3AsH_2 and the secondary arsine bis(trifuoromethyl) arsine $(CF_3)_2AsH$ are converted to gases at room temperature. All other arsines are liquids or solids. This must be kept in mind when treating such sample.

Notes:

Arsenate minerals

Arsenate minerals are those minerals containing the arsenate (AsO_4^{3-}) anion group. Both the Dana and the Strunz mineral classifications place the arsenates in with the phosphate minerals.

Aktashite is a rare arsenic sulfosalt mineral with formula $Cu_6Hg_3As_4S_{12}$. It is a copper mercury-bearing sulfosaltand is the only sulfosalt mineral with essential Cu and Hg yet known. It is of hydrothermal origin.

Adamite is a zinc arsenate hydroxide mineral, Zn_2AsO_4OH. It is a mineral that typically occurs in the oxidized or weathere dzone above zinc ore occurrences. Pure adamite is colorless, but usually it possess yellow color due to Fe compounds admixture.

Tints of green also occur and are connected with copper substitutions in the mineral structure. Olivenite is a copper arsenate that is isostructural with adamite and there is considerable substitution between zinc and copper resulting in an intermediate called *cuproadamite*. Zincolivenite is a recently discovered mineral being an intermediate mineral with formula $CuZn(AsO_4)(OH)$. Manganese, cobalt, and nickel also substitute in the structure. An analogous zinc phosphate, *tarbuttite*, is known.

Abernathyite is a rare hydrous uranium oxide mineral with chemical formula $K_2(UO_2)_2(AsO_4)_2 \cdot 6H_2O$. It occurs as a secondary fracture filling mineral. It crystallizes in the tetragonalsystem and forms yellow vitreous tabular crystals with perfect cleavage. It has a specific gravityof 3.3 and a hardness of 2.5.

The rare mineral **adelite**, is a calcium, magnesium, arsenate with chemical formula $CaMgAsO_4OH$. It forms a solid solution series with the vanadium bearing mineral gottlobite.

Various transition metals substitute for magnesium and lead replaces calcium leading to a variety of similar minerals in the *adelite - duftite group*.

Aerugite is a rare complex nickel arsenate mineral with a variably reported formula: $Ni_9(AsO_4)_2AsO_6$. It forms green to deep blue-green trigonalcrystals. It has a Mohs hardnessof 4 and a specific gravity of 5.85 to 5.95.

Agardite is a mineral group consisting of *agardite-(Ce), agardite-(Nd), agardite-(La), and agardite-(Y)*. They comprise a group of minerals that are hydrous hydrated arsenates of rare earth elements and copper, which contain variable amounts of calcium and sometimes lead. Yttrium, cerium, neodymium, lanthanum as well as trace to minor amounts of other rare earth elements are present in their structure. Agardite-(Y) is probably the most often found representative. The general formula for the group is $(REE,Ca)Cu_6(AsO_4)_3(OH)_6·3H_2O$. They form needle-like yellow-green (variably hued) crystals in the hexagonal crystal system.

Akrochordite is an exceptionally rare hydrated hydrous arsenate mineral of the formula $(Mn,Mg)_4(AsO_4)_2(OH)_4·4H_2O$ and represents a small group of rare in the nature manganese arsenates and, similarly to most other Mn-bearing arsenates, possess pinkish colour. It is typically associated with metamorphic Mn deposits.

Alarsite ($AlAsO_4$) is an aluminium arsenate mineral with its name derived from its composition: aluminium and arsenate. It occurs as brittle subhedral grains which exhibit trigonal symmetry.

It has a Mohs hardnessof 5-5.5 and a specific gravity of 3.32.

It is semitransparent, colorless with pale yellow tints and

shows a vitreous luster. It is optically uniaxial (+) with refractive indices of n_ω = 1.596 and n_ε = 1.608.

Cabalzarite $(Ca(Mg,Al,Fe^{3+})_2[AsO_4]_2 \cdot 20(H_2O,OH)$ is named for Walter Cabalzar of Switzerland.

Andyrobertsite is a rare, complex arsenate mineral with a blue color. It is named after Andrew C. Roberts, mineralogist with the Geological Survey of Canada. A Ca-rich analogue (with Ca instead of Cd) is called calcioandyrobertsite and has a more greenish tint.

Annabergite is an arsenate mineral consisting of a hydrous nickel arsenate, $Ni_3(AsO_4)_2 \cdot 8H_2O$, crystallizing in the monoclinic system and isomorphous with vivianite and erythrite. Crystals are minute and capillary and rarely met with, the mineral occurring usually as soft earthy masses and encrustations. A fine apple-green colour is its characteristic feature.

Arseniosiderite is a rare arsenate mineral formed by the oxidation of other arsenic-containing minerals, such as scorodite or arsenopyrite. It occurs in association with beudantite, carminite, dussertite, pharmacolite, pitticite, adamite and erythrite. The name arseniosiderite reflects two major elements of the mineral, arsenic and iron (Greek *sideros* means iron).

Arsenoclasite (originally **arsenoklasite**) is a red or dark orange brown mineral with formula $Mn_5(AsO_4)_2(OH)_4$. The name comes from the Greek words *αρσενικόν* (for arsenic) and *κλάσις* (for cleavage), as arsenoclasite contains arsenic and has perfect cleavage. Arsenoclasite is red or dark orange brown in color.

Arthurite is a mixture of divalent copper and iron ions in combination with trivalent arsenate, phosphate and sulfate ions with hydrogen and oxygen. Initially discovered by Arthur Russell in 1954. Arthurite is formed as a resultant mineral in the oxidation region of some copper deposits by the variation of enargite or arsenopyrite. The chemical formula of Arthurite is $CuFe_2^{3+}(AsO_4,PO_4,SO_4)_2(O,OH)_2 \cdot 4H_2O$.

Austinite is a member of the adelite-descloizitegroup, adelite subgroup, the Zn end member of the Cu-Zn series with conichalcite. It is the zinc analogue of cobaltaustinite and nickelaustinite. At one time "brickerite" was thought to be a different species, but it is now considered to be identical to austinite. Austinite is named in honour of Austin Flint Rogers, a mineralogist from Stanford University. The structure is composed of chains of edge-sharing polyhedra ZnO_6, and very distorted $Ca(O,OH)_8$ polyhedra linked through AsO_4 into a three-dimensional network.

Bayldonite (BAIL-done-ite) is a rare secondary mineral with the chemical formula $PbCu_3(AsO_4)_2(OH)_2H_2O$. It is named after its discoverer, John Bayldon

Beudandite is a secondary mineral occurring in the oxidized zones of polymetallic deposits. It is a lead, iron, arsenate, sulfate with end member formula: $PbFe_3(OH)_6SO_4AsO_4$.

Beudantite is in a subgroup of the alunitegroup. It is the arsenate analogue of the phosphate corkite. Beudantite also forms a solid-solution with segnitite and plumbojarosite.

It crystallizes in the trigonal crystal system and shows a variety of crystal habitsincluding tabular, acute rhombohedral, pseudo-cubic and pseudo-cuboctahedral.

Brassite is a mineralwith the chemical formula $Mg(AsO_3OH)\cdot 4(H_2O)$. It is white and leaves a white streak. It has perfect cleavage. Its crystals are orthorhombic- dipyramidal.

Bukovskyite (also known as "*clay of Kutná Hora*") is an iron arsenate sulfate mineralwith formula: $Fe_2(AsO_4)(SO_4)(OH)\cdot 7H_2O$ which forms nodules with a reniform (kidney-shaped) surface. Under a microscope, these nodules appear as a collection of minute needles similar to gypsum.

Some can be seen with the naked eye and occur inside the nodules.

Cahnite is a brittle white or colorless mineral that has perfect cleavage and is usually transparent. It usually forms tetragonal-shaped crystals and it has a hardness of 3 mohs. Cahnite was discovered in the year 1921. It was named Cahnite to honor Lazard Cahn.

The chemical formula for cahnite is $Ca_2B[AsO_4](OH)_4$. It is made up of 26.91% calcium, 3.63% boron, 25.15% arsenic, 1.35% hydrogen, and 42.96% oxygen. It has a molecular weight of 297.91 grams. Cahnite is not radioactive. Cahnite is associated with these other minerals: willemite, rhodonite, pyrochroite, hedyphane, datolite, and baryte.

Campylite is a variety of the lead arsenate mineral mimetite which received the name from the Greek 'kampylos'- bent, on account of the barrel-shaped bend of its crystals. It has also been used as an alternate name for pyromorphite

Chalcophyllite is a rare secondary copper arsenatemineral occurring in the oxidized zones of some arsenic-bearing copper deposits. A specimen is exhibiting partial replacement of liriconite, $Cu_2Al(AsO_4)(OH)_4 \cdot 4(H_2O)$, by chalcophyllite was documented.

The mineral is named from the Greek, *chalco*"copper" and *fyllon*, "leaf", in allusion to its composition and platy structure. It is a classic Cornishmineral that can be confused with tabular spangolite.

Clinoclase is a hydrous copper arsenate mineral, $Cu_3AsO_4(OH)_3$.

Conichalcite, $CaCu(AsO_4)(OH)$, is a relatively common arsenate mineral related to duftite ($PbCu(AsO_4)(OH)$). It is green, often botryoidal, and occurs in the oxidation zone of some metal deposits. It occurs with limonite, malachite, beudantite, adamite, cuproadamite, olivenite and smithsonite.

Cornubite is a rare secondary copper arsenate mineral with formula: $Cu_5(AsO_4)_2(OH)_4$.

Duftite is a relatively common arsenate mineral with the formula $CuPb(AsO_4)(OH)$, related to conichalcite. It is green and often forms botryoidal aggregates. It is a member of the Adelite-DescloiziteGroup, Conichalcite-Duftite Series. Duftite and conichalcite specimens from Tsumeb are commonly zoned in colour and composition. Microprobe analyses and X-ray powder-diffraction studies indicate extensive substitution of Zn for Cu, and Ca for Pb in the duftite structure. This indicates a solid solution among conichalcite, $CaCu(AsO_4)(OH)$, austinite, $CaZn(AsO_4)(OH)$ and duftite $PbCu(AsO_4)(OH)$, all of them belonging to the adelite group of arsenates.

Erythrite or **red cobalt** is a secondary hydrated cobalt arsenate mineral with the formula $(Co_3(AsO_4)_2 \cdot 8H_2O)$. Erythrite and annabergite $(Ni_3(AsO_4)_2 \cdot 8H_2O)$ (nickel arsenate) form a complete series with the general formula $(Co,Ni)_3(AsO_4)_2 \cdot 8H_2O$.

Euchroite is a hydrated copper arsenate hydroxide mineral with formula: $Cu_2AsO_4OH \cdot 3H_2O$. It is a vitreous green to emerald green mineral crystallizing in the orthorhombic system

Eveite is a manganese arsenate mineral in the olivenite group. Its chemical formula is Mn_2AsO_4OH. It is a dimorph of sarkinite and is isostructural with adamite. The name, for the biblical "Eve", comes from its structural similarities to adamite.

Eveite is anisotropic, which means that its physical and optical properties differ with respect to direction. It has high relief, which is the apparent topography exhibited by minerals in thin section as a consequence of refractive index. It is biaxial, so it has two optic axes and three indices of refraction n depending on the crystallographic direction. The refractive index is the ratio of the velocity of light in a vacuum to that in the mineral. The difference between the highest and lowest indices of refraction is called the birefringence, so the birefringence of eveite is $\beta = .032$.

Eveite is significant because it was the first mineral to show Mn^{2+} atoms in five-fold coordination, which is otherwise undocumented in mineral structures. It's an important addition to the olivenite group, and isostructural with andalusite. Because it shows up in very small quantities and in only two locations, it has no commercial use. It is relatively low-density and is associated with high hydrate and low density arsenates in open cavities, which contributes to its rarity.

Fluckite is a mineral with the chemical formula $CaMnH_2(AsO_4)_2·2H_2O$ It is named after the mineralogist Pierre Fluck of Louis Pasteur University in Strasbourg, France. The mineral contains arsenic in the form of arsenate $HAsO_4As(V)$. Arsenic has a myriad of uses "metallurgy, wood preservation, painting, medicine, pest control, and as an additive to chicken feed, where it increases growth," as well as being a strong and common poison. This form of arsenate is often found in ground water from deep wells and is a toxic substance.

Fornacite was first described in 1915 and named after Lucien Lewis Forneau. It is a rare lead, copper chromate arsenate hydroxide mineral with the formula: $Pb_2Cu(CrO_4)(AsO_4)(OH)$. It forms variably green to yellow, translucent to transparent crystals in the monoclinic- prismatic crystal system. It has a Mohs hardness of 2.3 and a specific gravity of 6.27.

Geigerite is a mineral, a complex hydrous manganese arsenate with formula: $Mn_5(AsO_3OH)_2(AsO_4)_2·10H_2O$. It forms triclinic pinacoidal vitreous colorless, red to brown crystals. It has a Mohs hardness of 3 and a specific gravity of 3.05.

Haidingerite is a calcium arsenate mineral with formula $Ca(AsO_3OH)·H_2O$. It crystallizes in the orthorhombic crystal system as short prismatic to equant crystals. It typically occurs as scaly, botryoidal or fibrous coatings. It is soft, Mohs hardness of 2 to 2.5, and has a specific gravity of 2.95. It has refractive indices of $n\alpha = 1.590$, $n\beta = 1.602$ and $n\gamma = 1.638$.

Jarosewichite is a rare manganese arsenate mineral with the formula: $Mn^{2+}_3Mn^{3+}(AsO_4)(OH)_6$. It was first described in Franklin, New Jerseywhich is its only reported occurrence. Its chemical composition and structure are similar to chlorophoenicite. This mineral is orthorhombic with 2/m2/m2/m point group. Its crystals are prismatic or barrel-shaped. The color of jarosewichite is dark red to black. It has subvitreous luster of fracture surfaces and reddish orange streak. This mineral occurs with flinkite, franklinite, andradite and cahnite.

The chemical composition of jarosewichite was obtained in 1982. These data were obtained by electron microprobe analysis with a voltage of 15 kV and a current of 0.025μA. Manganite(Mn), synthetic olivenite(As), synthetic ZnO(Zn), and hornblende(Ca, Mg, Fe) are used as standards for the analysis. Water percentage of the sample cannot be measured directly because of lacking large size of sample. The composition of jarosewichite is as follows:

- As_2O_5 24.0
- Mn_2O_3 17.7
- FeO 0.4
- MnO 42.3
- ZnO 1.2
- MgO 2.1
- CaO 0.2
- H_2O 12.1
- Total 100.0

The final calculation formula of unit cell contents is :
[$Mn^{3+}_{1.00}(Mn^{2+}_{2.74}Mg_{0.24}Fe_{0.03}Ca_{0.02}Zn_{0.07})_{\Sigma 3.10}(AsO_4)_{0.95}(OH)_{6.35}$], with Z=8 This result is very similar to the theoretical formula, which is $Mn^{2+}_3Mn^{3+}(AsO_4)(OH)_6$. The theoretical weight percent

of oxides are: Mn_2O_3=17.14, MnO=46.20, As_2O_5=24.95 and H_2O=11.71, and the sum is 100.

Kaatialaite ($Fe(H_2AsO_4)_3 \cdot 5H_2O$) is a ferric arsenate mineral.

Kankite is a mineral with the chemical formula $Fe^{3+}AsO_4 \cdot 3.5(H_2O)$. Kankite is a monoclinic mineral, meaning it is a mineral system having 3 unequal axes of which one is at right angles with the other two. It has an uneven fracture and has a hardness of 2-3 (gypsum-calcite). It is translucent yellowish-green in color with a grayish yellow streak. It's luster is dull to vitreous. Kankite contains the elements arsenic, iron, hydrogen and oxygen.

Keyite is a mineral with the chemical formula $Cu2+3Zn\ 4Cd\ 2(AsO4)_6 \cdot 2H2O$ named after Charles Locke Key, who is still live

Lavendulan is an uncommon copper arsenate mineral, known for its characteristic intense electric blue colour. It belongs to the lavendulan group, which has four members:

- Lavendulan $NaCaCu_5(AsO_4)4Cl.5H_2O$
- Lemanskiite $NaCaCu_5(AsO_4)4Cl.5H_2O$
- Sampleite $NaCaCu_5(PO_4)4Cl.5H_2O$
- Zdeněkite $NaPbCu_5(AsO_4)4Cl.5H_2O$

Lemanskiite and lavendulan are dimorphs; they have the same formula, but different structures. Lemanskiite is tetragonal, but lavendulan is monoclinic. Lavendulan has the same structure as sampleite, and the two minerals form a series. It is the calcium analogue of zdeněkite, and the arsenate analogue of sampleite.

Legrandite is a rare zinc arsenate mineral, $Zn_2(AsO_4)(OH) \cdot (H_2O)$.

It is an uncommon secondary mineral in the oxidized zone of arsenic bearing zinc deposits and occurs rarely in granite pegmatite. Associated minerals include: adamite, paradamite, kottigite, scorodite, smithsonite, leiteite, renierite, pharmacosiderite, aurichalcite, siderite, goethite and pyrite. It has been reported from Tsumeb, Namibia; the Ojuela mine in Durango, Mexico and at Sterling Hill, New Jersey, USA.

Liroconite is a complex mineral: Hydrated copper aluminium arsenate hydroxide, with the formula $Cu_2Al[(OH)_4|AsO_4]\cdot 4(H_2O)$. It is a vitreous monoclinic mineral, colored bright blue to green, often associated with malachite, azurite, olivenite, and clinoclase. It is quite soft, with a Mohs hardnessof 2 - 2.5, and has a specific gravity of 2.9 - 3.0.

Metazeunerite is an arsenate mineral with a chemical formula of $Cu(UO_2)_2(AsO_4)_2\cdot 8H_2O$. The origin of this mineral comes from the dehydrationprocess that metazeunerite must go through, and its association with zeunerite. As dehydration occurs, zeunerite loses an electron and is then metamorphosed into metazeunerite. Its crystal system is tetragonal and its crystal class is 4/m, which is also called the tetragonal-dipyramidal class because it only has a vertical four-fold rotation axis that is perpendicular to the symmetry plane. When looking at a thin section, metzeunerite is anisotropic, meaning that it has pleochroism. When a mineral is anisotropic, one can see whether it is unaixial or biaxial, depending on how fast the rays of light are moving through the mineral. This mineral is uniaxial negative due to the ordinary ray being slower than the extraordinary ray.

Mimetite, whose name derives from the Greek Μιμητής *mimethes*, meaning "imitator", is a lead arsenate chloride mineral ($Pb_5(AsO_4)_3Cl$) which forms as a secondary mineral in lead deposits, usually by the oxidation of galena and arsenopyrite. The name is a reference to mimetite's resemblance to the mineral pyromorphite. This resemblance is not coincidental, as mimetite forms a mineral series with pyromorphite ($Pb_5(PO_4)_3Cl$) and with vanadinite ($Pb_5(VO_4)_3Cl$).

Mixite is a rare copper bismuth arsenate mineral with formula: $BiCu_6(AsO_4)_3(OH)_6\cdot 3(H_2O)$. It crystallizes in the hexagonal crystal systemtypically occurring as radiating acicular prisms and massive encrustations. The color varies from white to various shades of green and blue. It has a Mohs hardness of 3.5 to 4 and a specific gravity of 3.8. It has an uneven fracture and a brilliant to adamantine luster.

It occurs as a secondary mineral in the oxidized zones of copper deposits. Associated minerals include: bismutite, smaltite, native bismuth, atelestite, erythrite, malachite and barite.

Olivenite is a copper arsenate mineral, formula Cu_2AsO_4OH. It crystallizes in the monoclinic system (pseudo-orthorhombic), and is sometimes found in small brilliant crystals of simple prismatic habitterminated by domal faces. More commonly, it occurs as globular aggregates of acicular crystals, these fibrous forms often having a velvety lustre; sometimes it is lamellar in structure, or soft and earthy.

A characteristic feature, and one to which the name alludes (German, *Olivenerz*, of A. G. Werner, 1789), is the olive-green color, which varies in shade from blackish-green in the crystals to almost white in the finely fibrous variety known as *woodcopper*. The hardness is 3, and the specific gravity is 4.3.

The arsenic of olivenite is sometimes partly replaced by a small amount of phosphorus, and in the species libethenitewe have the corresponding copper phosphate Cu_2PO_4OH.. Other members of this isomorphous group of minerals are adamite, Zn_2AsO_4OH, and eveite, Mn_2AsO_4OH.

Pharmacosiderite is a hydrated basic ferric arsenate, with chemical formula $KFe_4(AsO_4)_3(OH)_4 \cdot (6\text{-}7)H_2O$ and a molecular weightof 873.38 g/mol. It consists of the elements arsenic, iron, hydrogen, potassium, sodium and oxygen.

It has a Mohs hardnessof 2 to 3, about that of a finger nail. Its specific gravity is about 2.7 to 2.9, has indistinct cleavage, and is usually transparent or translucent. It has a yellow or white streak and a yellow, green, brown or red color. Its lustre is adamantine, vitreous and resinous, and it has conchoidal, brittle and sectile fracture.

Picropharmacolite, $Ca_4Mg(AsO_3OH)_2(AsO_4)_2 \cdot 11H_2O$, is a rare arsenate mineral. It was named in 1819 from the Greek for bitter, in allusion to its magnesium content, and its chemical similarity to pharmacolite.

The mineral irhtemite, $Ca_4Mg(AsO_3OH)2(AsO_4)_2 \cdot 4H_2O$, has the same composition as picropharmacolite, except that it has only four water molecules per formula unit, instead of eleven. It may be formed by the dehydrationof picropharmacolite.

Infrared spectra show that picropharmacolite contains water molecules H_2O, hydroxyl groups (OH)⁻ co-ordinated with Mg^{2+} cations, and acid arsenate radicals $(HAsO_4)^{2-}$. There are strong structural similarities with guerinite, $Ca_5(AsO_3OH)_2(AsO_4)_2 \cdot 9H_2O$ which indicates a similar formula for the two minerals.

X-ray diffraction methods indicate that As, Ca and Mg cations are positioned in corrugated layers parallel to the c axis, the layers being linked by hydrogen bonding only. Four independent water molecules are sandwiched between adjacent layers, and build up hydrogen-bonded chains which are also parallel to the c axis. The ratio of four Ca to one Mg remains fairly steady, and no significant Ca/Mg substitution occurs in any cation site. Hence if the formula of picropharmacolite is written as $Ca_4Mg(H_2O)_7(AsO_3OH)_2(AsO_4)_2 \cdot 4H_2O$[7], it is a better representation of the structure than the more usual formula $Ca_4Mg(AsO_3OH)_2(AsO_4)_2 \cdot 11H_2O$.

Roselite is a rare arsenate mineral with chemical formula: $Ca_2(Co,Mg)[AsO_4]_2 \cdot H_2O$.

Sewardite is a rare arsenate mineral with formula of $CaFe_2^{+3}(AsO_4)_2(OH)_2$

Sarkinite, synonymous with **chondrarsenite** and **polyarsenite**, is a mineral with formula $Mn_2(AsO_4)(OH)$.

Scorodite is a common hydrated iron arsenatemineral, with the chemical formula $FeAsO_4·2H_2O$.

Strashimirite is a rare monoclinic mineral containing arsenic, copper, hydrogen, and oxygen. It has the chemical formula $Cu_8(AsO_4)_4(OH)_4·5(H_2O)$.

Sewardite is a rare arsenate mineral with formula of $CaFe_2^{+3}(AsO_4)_2(OH)_2$

Strashimirite is a rare monoclinic mineral containing arsenic, copper, hydrogen, and oxygen. It has the chemical formula $Cu_8(AsO_4)_4(OH)_4·5(H_2O)$.

Tsumcorite is a rare hydrated lead arsenate mineral

Tyrolite is a hydrated calcium copper arsenate carbonate mineral with formula: $CaCu_5(AsO_4)_2CO_3(OH)_4·6H_2O$. Tyrolite forms glassy blue to green orthorhombicradial crystals and botryoidal masses. It has a Mohs hardnessof 1.5 to 2 and a specific gravity of 3.1 to 3.2. It is translucent with refractive indicesof $n\alpha=1.694$ $n\beta=1.726$ and $n\gamma=1.730$.

Vladimirite is a rare calcium arsenate mineral with a formulaof $Ca_5(HAsO_4)_2(AsO_4)_2·5H_2O$. It is named after the Vladimirovskoye deposit in Russia, where it was discovered in the 1950s.

Warikahnite is a zinc arsenate mineral of the triclinic crystal system.

Weilite (CaHAsO$_4$) is a rare arsenate mineral

Xanthiosite is an arsenate mineral

Zeunerite is a green copper uranium arsenate mineral with formula Cu(UO$_2$)$_2$(AsO$_4$)$_2$•10-16(H$_2$O). It is a member of the autunite group. Metazeunerite is a dehydration product.

Notes:

Arsenide minerals

An **arsenide mineral** is a mineral that contains arsenide as its main anion. Arsenides are grouped with the sulfidesin both the Dana and Strunz mineral classification systems

Examples

- algodonite Cu_6As
- domeykite Cu_3As
- löllingite $FeAs_2$
- nickeline $NiAs$
- rammelsbergite $NiAs_2$
- safflorite $(Co,Fe)As_2$
- skutterudite $(Co,Ni)As_3$
- sperrylite $PtAs_2$

The following notes will assist the bench chemist in recognizing compounds he may face during his wet-chemistry work.

Arsenolite is an arsenic mineral, chemical formula As_2O_3. It is formed as an oxidation product of arsenic sulfides. Commonly found as small octahedra it is white, but impurities of realgar or orpiment may give it a pink or yellow hue. It can be associated with its dimorph claudetite (a monoclinic form of As_2O_3) as well as realgar(As_4S_4), orpiment (As_2S_3) and erythrite, $Co_3(AsO_4)_2 \cdot 8H_2O$.

Arsenopyrite is an iron arsenic sulfide (FeAsS). It is a hard (Mohs 5.5-6) metallic, opaque, steel grey to silver white mineral with a relatively high specific gravity of 6.1. When dissolved in nitric acid, it releases elemental sulfur. When arsenopyrite is heated, it becomes magnetic and gives off toxic fumes. With 46% arsenic content, arsenopyrite, along with orpiment, is a principal oreof arsenic. When deposits of arsenopyrite become exposed to the atmosphere, usually due to mining, the mineral will slowly oxidize, converting the arsenic into oxides that are more soluble in water, leading to acid mine drainage.

Aktashite is a rare arsenic sulfosalt mineral with formula $Cu_6Hg_3As_4S_{12}$. It is a copper mercury-bearing sulfosalt and is the only sulfosalt mineral with essential Cu and Hg yet known. It is of hydrothermalorigin.

Alacránite(As_8S_9) is an arsenic sulfide mineral. It has a yellow-orange streak with a hardness of 1.5. It crystallizes in the monoclinic crystal system. It occurs with realgar and uzonite as flattened and prismaticgrains up to 0.5 mm across.

Arsenite minerals are very rare oxygen-bearing arsenic minerals.. The most often reported arsenite anionin minerals is the AsO_3^{3-} anion, present for example in reinerite $Zn_3(AsO_3)_2$. Unique diarsenite anions occur i. e. in leiteite $Zn[As_2O_4]$ and paulmooreite $Pb[As_2O_5]$. More complex arsenites include schneiderhöhnite $Fe^{2+}Fe^{3+}_3[As_5O_{13}]$ and ludlockite $PbFe^{3+}_4As_{10}O_{22}$.

Baumhauerite ($Pb_3As_4S_9$) is a lead sulfosalt mineral

Bellite is a mixture of minerals from Tasmaniawhich forms attractive orange red crystals. Its empirical formula, $PbCrO_4, AsO_4, SiO_2$ suggests a rather complex mixture of chromates and silicates of lead and arsenic.

Cervandonite is a rare arseonosilicate mineral, it has a chemical formula $(Ce,Nd,La)(Fe^{3+},Fe^{2+},Ti^{4+},Al)_3SiAs(Si,As)O_{13}$ or $(Ce,Nd,La)(Fe^{3+},Fe^{2+},Ti,Al)_3O_2(Si_2O_7)(As^{3+}O_3)(OH)$. It has a monocliniccrustal structure with supercell (Z=6), the crystal structure was established as a trigonal subcell, with space groupR3m and a = 6.508(1)Å, c = 18.520(3) Å, V 679.4(2) Å^3, and Z=3.

Christite is a mineral with the chemical formula $TlHgAsS_3$

Claudetite is an arsenic mineral chemical formula As_2O_3. It is named for the French chemist F. Claudet. Claudetite is formed as an oxidation product of arsenic sulfides and is colorless or white. It can be associated with arsenolite (the cubic form of As_2O_3) as well as realgar (As_4S_4), orpiment(As_2S_3) and sulfur.

Cobaltite is a sulfosalt mineral composed of cobalt, arsenic and sulfur, CoAsS. It contains up to 10 percent iron and variable amounts of nickel.[3] Structurally it resembles pyrite (FeS_2) with one of the sulfur atoms replaced by an arsenic atom.

Dimorphite (chemical name tetraarsenic trisulfide) is a very rare orange-yellow chalcogenide mineral. In nature, dimorphite forms primarily by deposit in volcanic fumaroles at temperatures of 70°-80°C (158°F-176°F).

Fettelite is a mercury-sulfosalt mineral with the chemical formula $Ag_{16}HgAs_4S_{15}$.

Gabrielite is a rare thallium sulfosalt mineral with a chemical formula of $Tl_6Ag_3Cu_6(As,Sb)_9S_{21}$[1] or $Tl_2AgCu_2As_3S_7$.

Galkhaite is a rare and chemically complex sulfosalt mineral from a group of natural thioarsenites. Its formula is $(Cs,Tl)(Hg,Cu,Zn)_6(As,Sb)_4S_{12}$, making the mineral the only known natural Cs-Hg and Cs-As phase. It occurs in Carlin-type hydrothermal deposits.

Geocronite is a mineral, a mixed sulfosalt containing lead, antimony, and arsenic with a formula of $Pb_{14}(Sb, As)_6S_{23}$.

Gersdorffite is a nickel arsenic sulfide mineral with formula NiAsS. It crystallizes in the isometric system showing diploidal symmetry. It occurs as euhedral to massive opaque, metallic grey-black to silver white forms. Gersdorffite belongs to a solid solution series with cobaltite, CoAsS. Antimony freely substitutes also leading to ullmannite, NiSbS. It has a Mohs hardnessof 5.5 and a specific gravity of 5.9 to 6.33.

Glaucodot is a cobalt iron arsenic sulfide mineral with formula: (Co,Fe)AsS. The cobalt:iron(II) ratio is typically 3:1 with minor nickel substituting. It forms a series with arsenopyrite (FeAsS). It is an opaque grey to tin-white typically found as massive forms without external crystal form.

Gratonite is a lead-arsenic sulfosalt mineral, with the chemical composition $Pb_9As_4S_{15}$.

Hutchinsonite is a sulfosalt mineral of thallium, arsenic and lead with formula $(Tl,Pb)_2As_5S_9$. Hutchinsonite is a rare hydrothermal mineral.

Jolliffeite is a selenide mineral with formula NiAsSe or (Ni,Co)AsSe. It is the selenium analogue of the sulfide mineral gersdorffite,

Jordanite is a sulfosalt mineral with chemical formula $Pb_{14}(As,Sb)_6S_{23}$ in the monoclinic crystal system.

Lorandite is a thallium arsenic sulfosalt with the chemical formula: $TlAsS_2$.

The mineral is being used for detection of solar neutrino via a certain nuclear reaction involving thallium. It has a monoclinic crystal structure consisting of spiral chains of AsS_3 tetrahedra interconnected by thallium atoms, and can be synthesized in the laboratory.

Madocite is a mineral with a general formula of $Pb_{17}(Sb,As)_{16}S_{41}$.

Madocite is anisotropic and classified as having high relief. It also displays strong pleochroism.

Marrite is depicted by the chemical formula $PbAgAsS_3$. It is the arsenic equivalent of mineral freieslebenite ($PbAgSbS_3$), but also displays close polyhedral characteristics with sicherite and diaphorite

Orpiment, As_2S_3, is a common monoclinic arsenic sulfide mineral. It has a Mohs hardness of 1.5 to 2 and a specific gravity of 3.49. It melts at 300 °C to 325 °C. Optically it is biaxial (−) with refractive indices of a=2.4, b=2.81, g=3.02.

Pääkkönenite is a metallic grey mineral with the molecular formula Sb_2AsS_2.

Pararealgar is an arsenic sulfide mineral with the chemical formula As_4S_4 also represented as AsS. It forms gradually from realga runder exposure to light. Its name derives from the fact that its elemental composition is identical to realgar, As_4S_4. It is soft with a Mohs hardnessof 1 - 1.5, is yellow orange in colour, and its monoclinic prismatic crystals are very brittle, easily crumbling to powder.

It is one of the sulfidesof arsenic and is one of two isomersof As_4S_4. It forms upon exposure of the symmetrical isomer to light. Its name derives from the fact that its elemental composition is identical to realgar, As_4S_4.

Structure

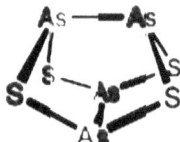 

The two isomers of As_4S_4:
realgar (left) and pararealgar.

Both isomers of As_4S_4 are molecular, in contrast to the other main sulfide of arsenic, orpiment(As_2S_3), which is polymeric. In pararealgar, there are three kinds of As centres (and three kinds of S centres). The molecule has C_s symmetry. In realgar, the four As (and four S) centres are equivalent and the molecule has D_{2d} symmetry. An analogous pair of isomers is also recognized for the corresponding phosphorus sulfides P_4S_4.

Pearceite is one of the four so-called "ruby silvers", pearceite $Cu(Ag,Cu)_6Ag_9As_2S_{11}$, pyrargyrite Ag_3SbS_3, proustite Ag_3AsS_3 and miargyrite $AgSbS_2$. It was discovered in 1896 and named after chemist Dr Richard Pearce. Originally pearceite was thought to be an arsenic analogue of polybasite

$Cu(Ag,Cu)_6Ag_9Sb2S_{11}$, and was called arsenpolybasite, and one polytypeof polybasite was called antimonpearceite.

Polybasite is a sulfosalt mineral of silver, copper, antimony and arsenic. Its chemical formula is $[(Ag,Cu)_6(Sb,As)_2S_7][Ag_9CuS_4]$.

Realgar, α-As_4S_4, is an arsenic sulfide mineral. It is a soft, sectile mineral occurring in monoclinic crystals, or in granular, compact, or powdery form, often in association with the related mineral, orpiment(As_2S_3).

Reinerite is a rare arsenite (arsenate(III)) mineral with chemical formula $Zn_3(AsO_3)_2$. It crystallizes in the orthorhombic crystal system.

Routhierite is a rare thallium sulfosalt mineral with formula $Tl(Cu,Ag)(Hg,Zn)_2(As,Sb)_2S_6$.

Seligmanite is a rare mineral, with the chemical formula $PbCuAsS_3$.

Stibarsen or **allemontite** is a natural form of arsenic antimonide (AsSb) or antimony arsenide (SbAs).

Teruggite is a mineral with the chemical formula $Ca_4MgAs_2B_{12}O_{22}(OH)_{12} \cdot 12(H_2O)$. It is colorless. Its crystals are monoclinicprismatic

Wakabayashilite is a rare arsenic, antimony sulfide mineralwith formula $[(As,Sb)_6S_9][As_4S_5]$.

Zimbabweite is a mineral; formula $(Na,K)_2PbAs_4(Nb,Ta,Ti)_4O_{18}$. It is generally classed as an arsenitebut is notable for also containing niobium and tantalum.

Zykaite or **zýkaite** is a grey-white mineral consisting of arsenic, hydrogen, iron, sulfur and oxygen with formula: $Fe^{3+}_4(AsO_4)_3(SO_4)(OH) \cdot 15(H_2O)$.

Roxarsone is a controversial arsenic compound used as a nutritional supplement for chickens

Arsenobetaine

Arsenic can be extracted as follows:

As2S3 + 6NaHS = 3H2S + As2S3.3Na2S

As2S3.3Na2S + 3H2 = 2As + 6NaHS

Arsenic has the characteristics of an è.ectronegative element. However, in certain its combination it may behave like an electropositive element, e.g. arsenic sulphate $As_2(SO_4)_3$. Arsenic replaces phosphorus and antimony in the oxides, chlorides and sulphides at 200° / 300°. Having said that, with the triphenyl compounds this replacement action is reversed – triphenyl stibine $Sb(C_6H_5)_3$ is decomposed by arsenic, and triphenylarsine $As(C_6H_5)_3$ is decomposed by phosphorus at 300°. You'll be able detect that garlic-like odour.

The author reported in 1994 the factor of vacuum and other treatments eliminating air. Neither hydrofluoric nor hydrochloric acid attacked arsenic in the absence of air, while in the presence of air trihalide is formed.

The formation of arsenic acid via alkali persulphate and the reaction with chlorosulphonic acid must be noted here as follows:

$$2As + 6HSO_3CL = 2AsCL_3 + 3SO_2 + 3H_2SO_4$$

In several environmental samples you may run into a red substance which is decomposed by water giving a violet black substance (which is Pas_4O_4). I found it to be a mixture of phosphorus trifluoride and arsenic. But, though phosphorus trichloride did not react with arsenic, if left over 10 hours in a temperature of about 200°, arsenic trichloride and phosorous are quantitatively formed. With phosphorus pentachloride the formation of arsenic and phosphorus trichlorides were observed. With phosphoryl chloride arsenic is dissolved.

Arsenic yields two series of salt, arsenious salts derived from arsenic trioxide and pentoxide.
The yellow flocculent precipitation you witness when treating hydrogen sulphide in acidic medium with arsenic trioxide, is arsenic trisulphide

$$2AsCL_3 + 3H_2S = 6HCL + As_2S_3$$

Sulphide is oxidized by conc. HNO3 to arsenic and sulphuric acids. It is also soluble in an ammoniacal soln. of hydrogen dioxide, and in a soln. of alkali hydroxide or sulphide, or in one ammonium carbonate. The reason behind this solubility is the formation of soluble alkali thioarsenites:

As2S3 + 6KH = 3H2O + As(OK)3 + As(SK)3

Or

As2S3 + 3(NH4)2S = 2As(SNH4)3

The reaction can be reversed by treating the solution with hydrochloric acid HCL. Hydrogen sulphide gives precipitation with no normal arsenites. The reason is the formation the soluble thioarenites:

As(OK)3 + 3H2S ⇌ 3 H2O + As(SK)3

Fluorine reacts violently with arsenic trioxide, forming a colourless liquid containing arsenic trifluoride and oxyfluoride.

A solution of arsenic trioxide in hydrochloric acid is oxidized by chlorine, and in the presence of alkali hydroxide, arsenate is formed.
The oxidation of arsenic trioxide by bromine, and the reduction of sodium hydroarsenate by hydrobromic acid is as follows:

As2Os + 4HBr   As2O3 + 4Br + 2H2O

A solution of arsenic trioxide reduces potassium peranganate and this reaction is accelerated in the presence of acids.

3As2O3 + 4KmnO4 = 3As2O5 + 2K2O + 4MnO2

"The end-point is reached when there is excess permanganate."

When copper sulphate is heated with aq. Ammonia and arsenic trioxide, in a sealed tube at 100°, it undergoes reduction to cuprous salt and arsenic acid if formed. The oxidation of cuprous soln. to the cupric condition by atm. Oxygen is as follows:

$2Cu'' + AsO_3''' + 2OH' = H_2O + 2Cu' + AsO_4'''$

the oxidation of the ammoniacal cuprous soln. to the cupric stage activates the atm. Oxygen for the oxidation of the residual unchanged arsenious acid.

The oxidation of sodium arsenite can be induced by the simultaneous oxidation of sodium sulphite, stannous chloride, manganous or cobaltous hydroxide and several aldehydes. Nitric oxide oxidizes sodium arsenite:

$2NO + Na_3AsO_3 = N_2O + Na_3AsO_4$

and with hydroxylamine, we have this reaction:

$NH_2OH + Na_3AsO_3 = Na_3AsO_4 + NH_3$

Another common fact you are likely to encounter in many industrial samples is arsnic reaction with aq. Soln. of copper arsenite:
$Cu_3(AsO_3)_2 + 2As = 3Cu + 2As_2O_3$

And, silver arsenite (insoluble in water) dissolves in acids and is soluble in the presence of alkali nitrate.

$(NH_4)_3AsO_3 + 2AgNO_3.NH_3 + H_2O =$
$\qquad (NH_4)_3AsO_4 + 2Ag + 2NH_4NO_3$

The following is a quick handbook note reference with the chemical structure of the arsenic compounds the bench chemist may need for analytical calculations:
Ammonium dihydroarsenate $NH_4H_2AsO_4$
Ammonium arsenatododecavanadatopentadecamolybdate

$6(NH_4)_2O \cdot AsO_5 \cdot 6V_2O_5 \cdot 15MoO_3 \cdot nH_2O$

Bismuthyl orthoarsenate $(BiO)_2AsO_4 \cdot H_2O$
Calcium triarsenatotetravanadate $2CaO \cdot 2V_2O_5 \cdot 3As_2O_5 \cdot 21H_2O$

Cupic triamminorthoarsenate $Cu_3(AsO_4)_2(NH_3)_3 \cdot 4H_2O$
Cupric diamminohydroarsenate $CuHAsO_4 \cdot 2NH_3 \cdot H_2O$

Copper hexahydroxyorthoarsenate $Cu_3(AsO_4)_2 \cdot 3Cu(OH)_2$
Calcium trioxyorthoarsenate $3CaO \cdot Ca(AsO_4)_2 \cdot 6H_2O$

Diammonium sodium orthoarsenate $(NH_4)_2NaAsO_4 \cdot 4H_2O$
Lead ferric hydroxysulpatophosphatarsenate
$4Fe_2O_3 \cdot 2PbO \cdot 3SO_3 \cdot (As,P)_2O_5 \cdot 9H_2O$

Nickel triarsenatotetravanadate $2NiO \cdot 2V_2O_5 \cdot 3As_2O_5 \cdot 24H_2O$
Potassium beryllium oxydiorthoarsenate $2KBeAsO_4 \cdot BeO \cdot 10H_2O$

Potassium arsenatohexavanadatopentadecamolybdate
$6K_2O \cdot As_2O_5 \cdot 3V_2O_5 \cdot 15MoO_3 \cdot nH_2O$
Potassium magnesium dihydrotriothoarsenate
$Mg_3KH_2(AsO_4)_3 \cdot 5H_2O$

Potassium tetrahydrodiarsenatoctodecatungstate
$K_6H_4\{As_2(OH)_2(W_2O_7)_9\} \cdot 11H_2O$
Sodium hydrotrioxysulpharsenate $Na_2HAsO_3S \cdot 8H_2O$
Zinc dihydroxypyroarsenate $Zn(OH)_2 \cdot Zn_2As_2O_7 \cdot 7H_2O$

Though you probably won't memorize all these facts, the use of this reference is clearly a useful approach. With nearly two decades of analytical chemistry, I still once in a while resort to my notes for a compound formula or characteristics.

Selenium

notes

Selenium is commonly found in soil samples as metal in association with sulfide ores, basic ferric selenite $Fe(OH)SeO_3$, ferric selenite and organo compounds. Selenium occurs in alkaline soils mainly as selenates.

In acid soils it exists mainly as selenides. In slimes it exists mainly as $CuAgSe$ and Ag_2Se. Based on sample matrix, the following reactions are found to be common:

The oxidation of selenides and its conversion to selenite:

Se(selenide) + Na_2CO_3 + O_2 → Na_2SeO_3 + CO_2

Se(selenide) + Na_2CO_3 + O_2 → Na_2SeO_4 + CO_2

Ferrous sulfate accelerates the reaction as follows:

**H_2SO_4 + HCL + $FeSO_4$ →
H_2SeO_3 + $FeCL_3$ + $Fe_2(SO_4)_3$ + H_2O**

Selenium usually exists in nature as metal selenides in association with sulfide ores. Elemental selenium, basic ferric selenite $Fe(OH)SeO_3$, calcium selenate and organoselenium compounds are also found in soil samples.

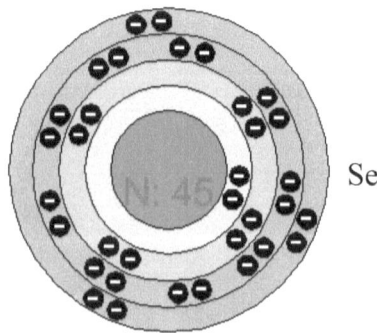

- Acid digestion is necessary to ensure that Se in the sample is in the inorganic form. Se^{VI} is not quantitatively recovered by hydride generation and must be reduced to Se^{IV}. This requires sample preparation in HCL {5-10 M} in 80-90 ^0C for 10-15 minutes.

- Other common selenium digestion reactions / complications / side effects:

With sodium sulfite:

$Se_2 + Na_2SO_3 \rightarrow Na_2SeO_3$
With sulfuric acid: **$Na_2SeSO_3 + H_2SO_4 \rightarrow Na_2SO_4 + Se + SO_2 + H_2O$**

- Trace level of potassium iodide will interfere severely with the determination of Se. Special Se tubes / cell / glass separator must be assigned strictly for Se analysis.

Selenium forms halides by reacting with fluorine and chlorine and to a lesser degree with terhalogen compounds and bromine. It decomposes hydrogen iodide to liberate iodine. Selenium dissolves in alkali-metal sulfites forming selenosulfates, M_2SSeO_3.

Selenium is oxidized by solutions of alkali-metal dichromates, permanganates, chlorates and calcium hypochlorite. It dissolves in strong alkaline solutions yielding selenides and selenites. It forms selenocyanates, MSeCN with many inorganic and organic derivatives of HSeCN.

The role of KI is also evident in its use in the thiosulfate titration as follows:

$$H_2SeO_3 + HCL + KI \rightarrow Se + I_2 + KCL + H_2O$$

$$H_2SeO_3 + Na_2S_2O_3 + HCL \rightarrow Na_2SeS_4O_6 + Na_2S_4O_6 + NaCL + H_2O$$

- Selenium combines directly with fluorine, chlorine and bromine, but not iodine, and forms the monohaliodes Se_2X_2, the dihalides SeX_2, the tetrahalides SeX_4 and the hexafluoride SeF_6. The compounds are covalent and volatile.

- Selenium oxyhalides $SeOX_2$ dissolves many metal chlorides and chalcogenides.

Common selenium acids and salts often present and active in the test tube include selenious acid H_2SeO_3, selenic acid H_2SeO_4, selenium oxyacids such as permonoselenic acid H_2SeO_5, perdiselenic acid $H_2Se_2O_8$, and Pyroselenic acid $H_2Se_2O_7$. Other commonly interfering inorganic selenium compounds include sodium selenocyanate NaSeCN, selenocyanogen SeCN / $(SeCN)_2$, and selenium selenosulfate Na_2SeSO_3.

- Selenium is found impurely in metal sulfide ores, where it partially replaces the sulfur. Commercially, selenium is produced as a byproduct in the refining of these ores, most often during copper production.

Selenium salts are toxic in large amounts, but trace amounts are necessary for cellular function in many organisms, including all animals. Selenium is a component of the antioxidant enzymes glutathione peroxidase and thioredoxin reductase (which indirectly reduce certain oxidized molecules in animals and some plants). It is also found in three deidinase enzymes, which convert one thyroid hormone to another. Selenium requirements in plants differ by species, with some plants requiring relatively large amounts, and others apparently requiring none.

- The most stable and dense form of selenium has a gray colour and hexagonal crystal lattice consisting of helical polymeric chains, wherein the Se-Se distance is 237.3 pm and Se-Se-Se angle is 130.1 degrees. The minimum distance between chains is 343.6 pm.

Gray Se is formed by mild heating of other allotropes, by slow cooling of molten Se, or by condensing Se vapors just below the melting point. Whereas other Se forms are insulators, gray Se is a semiconductor showing appreciable photoconductivity. Contrary to other allotropes, it is unsoluble in CS_2. It resists oxidation by air and is not attacked by non-oxidizing acids. With strong reducing agents, it forms polyselenides. Selenium does not exhibit the unusual changes in viscosity that sulfur undergoes when gradually heated

Selenium has six naturally Occurring isotopes, five of which are stable: ^{74}Se, ^{76}Se, ^{77}Se, ^{78}Se, and ^{80}Se.

Selenium compounds commonly exist in the oxidation states −2, +2, +4, and +6.

Chalcogen compounds

Selenium forms two oxides: selenium dioxide (SeO_2) and selenium trioxide (SeO_3). Selenium dioxide is formed by the reaction of elemental selenium with oxygen:

$Se_8 + 8\ O_2 \rightarrow 8\ SeO_2$

Structure of the polymer SeO_2. The (pyramidal) Se atoms are yellow.
It is a polymeric solid that forms monomeric SeO_2 molecules in the gas phase. It dissolves in water to form selenous acid, H_2SeO_3. Selenous acid can also be made directly by oxidizing elemental selenium with notric acid:

$3\ Se + 4\ HNO_3 + H_2O \rightarrow 3\ H_2SeO_3 + 4\ NO$

Unlike sulfur, which forms a stable trioxide, selenium trioxide is thermodynamically unstable and decomposes to the dioxide above 185 °C:

$2\ SeO_3 \rightarrow 2\ SeO_2 + O_2$ ($\Delta H = -54$ kJ/mol)

Selenium trioxide is produced in the laboratory by the reaction of anhydrous potassium selenate (K_2SeO_4) and sulfur trioxide (SO_3).

Salts of selenous acid are called selenites. These include silver selenite (Ag_2SeO_3) and sodium selenite (Na_2SeO_3).

Hydrogen sulfide reacts with aqueous selenous acid to produce selenium disulfide::

$$H_2SeO_3 + 2\ H_2S \rightarrow SeS_2 + 3\ H_2O$$

Selenium disulfide consists of 8-membered rings of a nearly statistical distribution of sulfur and selenium atoms. It has an approximate composition of SeS_2, with individual rings varying in composition, such as Se_4S_4 and Se_2S_6. Selenium disulfide has been use in shampoo as an anti-dandruff agent, an inhibitor in polymer chemistry, a glass dye, and a reducing agent in fireworks.

Selenium trioxide may be synthesized by dehydrating selenic acid, H_2SeO_4, which is itself produced by the oxidation of selenium dioxide with hydrogen peroxide:

$$SeO_2 + H_2O_2 \rightarrow H_2SeO_4$$

Hot, concentrated selenic acid is capable of dissolving gold, forming gold(III) selenate.

Halogen compounds

Iodides of selenium are not well known. The only stable chloride is selenium monochloride (Se_2Cl_2), which might be better known as selenium(I) chloride; the corresponding bromide is also known. These species are structurally analogous to the corresponding disulfur dichloride.

Selenium dichloride is an important reagent in the preparation of selenium compounds (e.g. the preparation of Se_7). It is prepared by treating selenium with sulfuryl chloride (SO_2Cl_2). Selenium reacts with fluorine to form selenium hexafluoride:

$Se_8 + 24 F_2 \rightarrow 8 SeF_6$

In comparison with its sulfur counterpart (sulfur hexafluoride), selenium hexafluoride (SeF_6) is more reactive and is a toxic pulmonary irritant. Some of the selenium oxyhalides, such as selenium oxyfluoride ($SeOF_2$) and selenium oxychloride ($SeOCl_2$) have been used as specialty solvents.

Selenides

Analogous to the behavior of other chalcogens, selenium forms a dihydride H_2Se. It is a strongly odiferous, toxic, and colorless gas. It is more acidic than H_2S. In solution it ionizes to $HSe-$. The selenide dianion Se^{2-} forms a variety of compounds, including the minerals from which selenium is obtained commercially. Illustrative selenides include mercury selenide (HgSe), lead selenide (PbSe), zinc selenide (ZnSe), and copper indium gallium diselenide ($Cu(Ga,In)Se_2$). These materials are semiconductors. With highly electropositive metals, such as aluminium, these selenides are prone to hydrolysis:

$Al_2Se_3 + 6 H_2O \rightarrow Al_2O_3 + 6 H_2Se$

Alkali metal selenides react with selenium to form polyselenides, Se_2-x, which exist as chains.

Other compounds

Tetraselenium tetranitride, Se_4N_4, is an explosive orange compound analogous to tetrasulfur tetranitride (S_4N_4). It can be synthesized by the reaction of selenium tetrachloride ($SeCl_4$) with $[(CH_3)_3Si)_2N]_2Se$.

Selenium reacts with cyanides to yield selenocyanates:
8 KCN + Se_8 → 8 KSeCN

Organoselenium compounds

Selenium, especially in the II oxidation state, forms stable bonds to carbon, which are structurally analogous to the corresponding organosulfur compounds. Especially common are selenides (R_2Se, analogues of thioethers), diselenides (R_2Se_2, analogues of disulfides), and selenols (RSeH, analogues of thiols). Representatives of selenides, diselenides, and selenols include respectively selenomethionine, diphenyldiselenide, and benzeneselenol.

The sulfoxidein sulfur chemistry is represented in selenium chemistry by the selenoxides (formula RSe(O)R), which are intermediates in organic synthesis, as illustrated by the selenoxide elimination reaction. Consistent with trends indicated by the double bond rule, selenoketones, R(C=Se)R, and selenaldehydes, R(C=Se)H, are rarely observed.

Biological role

Selenium is toxic in large doses, however, it is an essential micronutrient for mammals. In plants, it occurs as a bystander mineral, sometimes in toxic proportions in Selenium is a component of the unusual amino acids selenocysteine and selenomethionine. In humans, selenium is a trace element nutrient that functions as cofactor for reduction of antioxidant enzymes, such as glutathione peroxidases and certain forms of thioredoxin reductase found in animals and some plants (though this enzyme occurs in all living organisms, not all forms of it in plants require selenium).

The glutathione peroxidase family of enzymes (GSH-Px) catalyze certain reactions that remove reactive oxygen species such as hydrogen peroxide and organic hydroperoxides:

2 GSH + H_2O_2 ----GSH-Px → GSSG + 2 H_2O

Selenium also plays a role in the functioning of the thyroid gland and in every cell that uses thyroid hormone, by participating as a cofactor for the three of the four known types of thyroid hormone deiodinases, which activate and then deactivate various thyroid hormones and their metabolites (the iodothyronine deiodinases are the subfamily of deiodinase enzymes that use selenium as the otherwise rare amino acid selenocysteine.
Only the deiodinase iodotyrosine deiodinase, which works on the last break-down products of thyroid hormone, does not use selenium.)

Organoselenium compounds are chemical compounds containing carbon-to-selenium chemical bonds. Organoselenium chemistryis the corresponding science exploring their properties and reactivity. Selenium belongs with oxygen and sulfur to the group 16 elements and similarities in chemistry are to be expected.

Selenium can exist with oxidation state-2, +2, +4, +6. Se(II) is the dominant form in organo-selenium chemistry. Down the group 16 column, the bond strength becomes increasingly weaker (234 kJ/mol for the C–Se bond and 272 kJ/mol for the C–S bond) and the bond lengths longer (C–Se 198 pm, C–S 181 pm and C–O 141 pm).

Selenium compounds are more nucleophilic than the corresponding sulfur compounds and also more acidic. The pKa values of XH_2 are 16 for oxygen, 7 for sulfur and 3.8 for selenium. In contrast to sulfoxides, the corresponding selenoxides are unstable in the presence of β-protons and this property is utilized in many organic reactions of selenium, notably in selenoxide oxidations and in selenoxide eliminations.

Organo-selenium compounds are found at trace levels in ambient waters, soils and sediments. The first organo-selenium compound ever isolated was diethyl selenide back in 1836.

Structural classification of organoselenium compounds

Selenols (RSeH) are the selenium equivalents of alcohols and thiols. These compounds are relatively unstable and generally have an unpleasant smell. Phenylselenol (also called selenaphenol or PhSeH) is more acidic (pKa 5.9) than thiophenol (pKa 6.5) and also oxidizes more readily to the diselenide. Selenaphenol is prepared by reduction of diphenyldiselenide.

Diselenides (R-Se-Se-R) are the selenium equivalents of peroxides and disulfides. They are useful shelf-stable precursors to more reactive organo-selenium reagents such as selenols and selenenyl halides.

Diphenyldiselenide is prepared from phenylmagnesium bromideand selenium followed by oxidation of the product PhSeMgBr.

Selenenyl halides(R-Se-Cl, R-Se-Br) are prepared by halogenation of diselenides. Bromination of diphenyldiselenide gives phenylselenyl bromide (PhSeBr). These compounds are sources of "PhSe+".

Selenides(R-Se-R), also called selenoethers, are the selenium equivalents of ethers and thioethers. These are the most prevalent organo-selenium compounds.

Symmetrical selenides are usually prepared by alkylation of alkali metal selenide salts, e.g. sodium selenide. Unsymmetrical selenides are prepared by alkylation of selenoates.

These compounds are typically react as a nucleophiles, e.g. with alkyl halides (R'-X) to give selenonium salts R'RRSe+X-. Divalent selenium can also interact with soft heteroatoms to form hypervalent selenium centers.

They also react in some circumstances as electrophiles, e.g. with organolithium reagents (R'Li) to the ate complex R'RRSe-Li+.

Selenoxides(R-Se(O)-R) are the selenium equivalents of sulfoxides. They can be further oxidized to selenones R-Se(O)2R, the selenium analogues of sulfones.

Perseleninic acids (RSe(O)OOH) catalyse epoxidation reactions and Baeyer–Villiger oxidations.

Selenuranes are hypervalentorganoselenium compounds, formally derived from the tetrahalides such as SeCl4. Examples are of the type ArSeCl3.

The chlorides are obtained by chlorination of the selenenyl chloride.

Seleniranesare three-membered rings (parent: C2H4Se) related to thiiranesbut, unlike thiiranes, seleniranes are kinetically unstable, extruding selenium directly (without oxidation) to form alkenes. This property has been utilized in synthetic organic chemistry.

Selones (R2C=Se, sometimes called selenones) are the selenium analogues of ketones. They are rare because to their tendency to oligomerize.
Diselenobenzoquinone is stable as a metal complex
Selenoureais an example of a stable compound containing a C=Se bond.

Organo-selenium compounds in nature

Selenium, in the form of organo-selenium compounds, is an essential micronutrient whose absence from the diet causes cardiac muscle and skeletal dysfunction.
Organo-selenium compounds are required for cellular defense against oxidative damage and for the correct functioning of the immune system. They may also play a role in prevention of premature aging and cancer.

Most known selenium-dependent enzymes contain selenocysteine, a cysteineanalog in which selenium replaces sulfur and which is encoded in a special manner by DNA. Selenocysteine, called the twenty-first amino acid, is essential for ribosome-directed protein synthesis.

More than 25 selenium-containing proteins (selenoproteins) are now known. Selenomethionineis a selenide-containing amino acid that also occurs naturally, but is generated by post-transcriptional modification.

Glutathione oxidaseis an enzyme with a diselenide at its active site. Organoselenium compounds have been found in higher plants. For example, upon analysis of garlic using the technique of high-performance liquid chromatography combined with inductively coupled plasma mass spectrometry (HPLC-ICP-MS), it was found that γ-glutamyl-Se-methylselenocysteine was the major Se-containing component, along with lesser amounts of Se-methylselenocysteine. Trace quantities of dimethyl selenide and allyl methyl selenide are found in human breath after consuming raw garlic.

Vinylic selenides are organoselenium compounds that play a role in organic synthesis, especially in the development of convenient stereoselective routes to functionalized alkenes. Although various methods are mentioned for the preparation of vinylic selenides, a more useful procedure has centered on the nucleophilic or electrophilicorganoselenium addition to terminal or internal alkynes.

For example, the nucleophilic additionof selenophenol to alkynes affords, preferentially, the Z-vinylic selenides after longer reaction times at room temperature. The reaction is faster at a high temperature; however, the mixture of Z- and E-vinylic selenides was obtained in an almost 1:1 ratio. On the other hand, the adducts depend on the nature of the substituents at the triple bond. Conversely, vinylic selenides can be prepared by palladium-catalyzed hydroselenation of alkynesto afford the Markownikov adduct in good yields.

There are some limitations associated with the methodologies to prepare vinylic selenides illustrated above; the procedures described employ diorganoyl diselenides or selenophenolas starting materials, which are volatile and unstable and have an unpleasant odor. Also, the preparation of these compounds is complex.

Selenoxide oxidations

Selenium dioxideis useful in organic oxidation. Specifically, SeO2 will convert an allylic methylenegroup into the corresponding alcohol. A number of other reagents bring about this reaction. .

In terms of reaction mechanism, SeO2 and the allylic substrate react via pericyclicprocess activating the C-H bond. The second step is a sigmatropic reaction. Oxidations involving selenium dioxide are often carried out with catalytic amounts of the selenium compound and in presence of a sacrificial catalyst or co-oxidant such as hydrogen peroxide.

SeO2-based oxidations sometimes afford carbonyl compounds such as ketones., β-Pinene[24] and cyclohexanone oxidation to 1,2-cyclohexanedione

Oxidation of ketones having α-methylene groups affords diketones. This type of oxidation with selenium oxide is called Riley oxidation.

Selenoxide eliminations

In presence of a β-proton, a selenide will give an elimination reaction after oxidation, to leave behind an alkene and a selenol. In the elimination reaction, all five participating reaction centers are coplanar and, therefore, the reaction stereochemistry is syn. Oxidizing agents used are hydrogen peroxide, ozone or MCPBA. This reaction type is often used with ketones leading to enones. An example is acetylcyclohexanone elimination with benzeneselenylchloride and sodium hydride

Selenium may be recovered by roasting the mud with soda or sulfuric acid, or by smelting with soda and niter:
$Cu_2Se + Na_2CO_3 + 2O_2 \rightarrow 2CuO + Na_2SeO_3 + CO_2$

The selenite Na_2SeO_3 is acidified with sulphuric acid. Tellurites precipitate out of solution, leaving selenous acid, H_2SeO_3n. Selenium is liberated from selenous acid by SO_2

$H_2SeO_3 + 2SO_2 + H_2O \rightarrow Se + 2H_2SO$

Additional selenium bench notes

Selenium forms sodium selenide when treated with sodium hyposulphite. When selenium acts on potassium or ammonium hydrosulphite, this is what happens:

$$4MHSO_4 \longrightarrow 2M_2SO_4 + SO_2 + S + 2H_2O$$

The problem some bench analysts may experience is that the indicated reaction, doesn't actually occur as stated above, it happens in 3 stages as follows:

1) $M_2SO_3 + Se \longrightarrow M_2SSeO_3$

2) $M_2SSeO_3 + M_2SO_3 + 2H_2SO_3 \longrightarrow$
   $M_2SO_4 + M_2S_3O_6 + Se + 2H_2O$

3) $M_2S_3O_6 \longrightarrow M_2SO_4 + SO_2 + S$

Thus, the true catalytic action of the selenium ceases when only one-half of the total sulphuric acid if formed.

Carbon diselenide, ethyl sullphide and ethyl selenide, any of them would dissolve vitreous selenium, but the metallic selenium is insoluble in ethyl sulphide and in tetrachloride.

Metallic selenium is slightly soluble in toluene, quinoline and nitrobenzene. As explained in details in my experimental paper, in connection with the allotropic forms of selenium, vitreous selenium dissolves in benzene, acetophenone, acetone, isobutyric acid, alcohol, ethyl acetate, benzonitrile, thiophene, toluene, acetone and chloroform, forming red crystals; and in quinoline, aniline and pyridine, etc., forming metallic selenium. In potassium cyanide solution, selenium forms potassium selenocyanate.

One more caution here: during bench work, many chemists have assumed that the peculiar smell generated when selenium burns is selenium monoxide. I invite you to reconsider. A better explanation is the presence of traces of moisture forming hydrogen selenide. The characteristic radish-like odour of selenium when heated on charcoal is due to carbon diselenide and not to selenium suboxide.

When you need to introduce sulphur, free from sulphur dioxide, to a solution of hydrogen selenide or by bubbling the gas through water with sulphur in suspension, the sulphur is instantly coloured: at first redish orange, then deep red, then dark redish brown. Keep in mind now that you have formed hydrogen sulphide:

$H_2Se + S = Se + H_2S$

The reaction between hydrogen sulphide and selenious acid, and between sulphurous acid and hydrogen selenide are not as simple as they may theoretically be assumed to be. When a cold diluted solution of selenious acid is treated with hydrogen sulphide, a lemon-yellow, pulverulent precipitate is formed. This precipitation contains sulphur and selenium in the proportions 2:1. this is what happens:

$2H_2S + H_2SeO_3 \longrightarrow 2S + Se + 3H_2O$

if the precipitate if left for some time moistened with carbon disulphide, an orange-red layer or scales appear which is actually SeS, and the excess of sulpur is removed in the solvent. Now, if the solution of selenious acid is warm, the precipitate with hydrogen sulphide is plastic and red. Another reaction occurs. Sulphuric acid is formed and the proportion of sulphur in precipitate in less than 2:1. This is why:

$H_2S + 2H_2SeO_3 = H_2SO_4 + 2Se + 2H_2O$

Selenide is slowly dissolved by nitric acid and readily dissolved by aqua regia. Nitric acid converts the selenide into selenite.

Ammonia reacts with selenium dioxide evolving nitrogen with the separation of selenium.
Ammonia acts on a solution of selenium dioxide in absolute alcohol forming what many analysts still assume it is ammonium amide selenite, when in fact it is $NH_4(C_2H_5)SeO_3$ which makes it ammonium ethyl selenite.

Selenium dioxide is reduced when heated in hydrogen. The dioxide attracts moisture from the air. Selenium dioxide absorbs dry hydrogen fluoride forming oxyhydrofluorides. When distilled with sodium chloride, selenyl chloride is formed as follows:

$$2SeO_2 + 2NaCL = Na_2SeO_3 + SeOCL_2$$

selenium dioxide yields nitrogen and black, amorphous selenium when treated with hydrazine as follows:

$$SeO_3 + N_2H_4 = N_2 + Se + 2H_2O$$

And nitrogen and reddish-brown amorphous selenium with hydroxylamine hydrochloride as follows:

$$SeO_2 + 4NH_2OH = 2N_2 + Se + 6H_2O$$

By passing phosphine into an alcoholic solution of selenium dioxide, a pale yellow precipitate of phosphorus and selenium is formed while a great portion of the selenium remains in solution as ethyl selenide $(C_2H_5)_2Se$.
Selenium dioxide is converted by phosphorus pentachloride into selenium tetrachloride and phosphoryl chloride as follows:

$$SeO_2 + 2PCL_5 = SeCL_4 + 2POCL_3$$

And by phosphorus trichloride into brown amorphous selenium and phosphoryl chloride as follows:

$SeO_2 + 2PCL_2 = Se + 2POCL_3$

One common error analysts make is not considering the temperature factor in the reduction of selenic acid by hydrogen sulphide. The mere theoretical equation will not be totally accurate on the laboratory bench. The rate of reduction increases both with the temperature and the acid concentration. At 10% solution of selenic acid at 45° complete decomposition takes 13.5 hours in this fashion:

$H_2S + H_2SeO_4 = 2H_2O + SO_2 + Se$

And the sulphur dioxide then reacts with hydrogen sulphide depositing sulphur as follows:

$3H_2S + H_2SeO_4 = 4H_2O + Se + 3S$

The theoretical assumption that hydrogen sulphide and selenic acid will always yield selenium or selenium sulphide does not take into consideration that selenic acid – freed from selenious acid – after dilution to 5% or lower, can be subjected to a stream of hydrogen sulphide without showing the slightest trace of a yellow coloration in the flask.

$H_2SeO_4 + 3SO_2 + 2H_2O = 3H_2SO_4 + Se$

Some chemists didn't realize in the past is that this reaction may happen in two stages as follows:

$H_2SeO_4 + SO_2 = H_2SO_4 + SeO_2$
$SeO_2 + 2H_2O + 2SO_2 = Se + 2H_2SO_4$

A few rock and soil refinery sample analysis I ran into demonstrated the same issue of the importance of elucidating the intricacies of the reaction beyond the obvious theoretical assumption. They involved the fact that selenic acid dissolves Cu & Au but not Pt, forming selenious acid; and it dissolves Zn & Fe with the liberation of hydrogen. One must consider the possible effect of selenic acid as an impurity in the nitric acid. It was obvious that the formation of selenium was due to the reducing action of nascent hydrogen on selenic acid:

$6H + H_2SeO_4 = Se + 4H_2O$

the final result of this slow reaction therefore can be written as such:

$3Fe + 4H_2SeO_4 = 3FeSeO_4 + Se + 4H_2O$

when iron is replaced by zinc, hydrogen is given off, and reduction of selenic acid in drastically reduced.

Meanwhile, in the case of magnesium, hydrogen is also evolved, but solid reduction occurs.

When selenic acid is heated with excess Hg, mercury selenite that first appear:

$Hg + H_2SeO_4 = H_2O + HgSeO_3$

reacts with excess selenic acid, the mercury selenite reacts with the acid to produce mercury selenate.
When dealing with selenium halides, the preceding caution or advice is even more significant.

Monochloride decomposes when distilled:

$2Se_2CL_2 = SeCL_4 + 3Se$

Hydrogen has no action on selenium monochloride and will decompose in water into hydrogen chloride, selenium and selenic acid:

$2Se_2CL_2 + 2H_2O = SeO_2 + 3Se + 4HCL$

some sulphur in sulphur monochloride can be displaced by selenium:

$S_2Cl_2 + 2Se = Se_2Cl_2 + 2S$

Selenium oxychloride dissolves selenium, and selenium monochloride dissolves selenium dioxide:

$2Se_2Cl_2 + SeO_2 \rightarrow 2SeOCl_2 + 3Se$

When selenium tetrachloride is introduced to water, heat is emitted due to this fact:

$SeCl_4 + 2H_2O = 4HCl + SeO_2$

If the amount of water is minimum or mere moisture in the air (in the flask or test tube), selenium oxychloride is formed.
Liquid hydrogen sulphide reacts at low temperature with selenium tetrachloride, forming selenium monochloride and red selenium is observed in the flask. With sulphur dioxide no reaction was recorded on selenium tetrachloride. However, with sulphur trioxide a complex was formed $SO_3 \cdot SeCl_4$. and with sulphuric acid we see this reaction:

$3SeCl_4 + 2H_2SO_4 = SeO_2 + 2(SO_3 \cdot SeCl_4) + 4HCl$

while part reacts: $3SeCl_4 + 2H_2SO_4 = SeOCl_2 + H_2S_2O_7 + 2HCl$

Followed by: $SeOCl_2 + 2H_2SO_4 = SeO_2 + H_2S_2O_7 + 2HCl$

With pyrosulphuric acid the reaction is different:

$H_2S_2O_7 + SeCl_4 = H_2SO_4 + SO_3 \cdot SeCl_4$

And, when dealing with oxyhalogen compounds, it must be stated that not all compounds react according to a pattern, and, one must consider, on paper, all possible scenarios before approaching the test tube or beaker.

Selenium oxyfluride SeOF2 or selenyl fluoride, prepared by the action of dry silver fluoride on selenium oxychloride (in a platinum bottle) with subsequent distillation and condensation in a platinum condenser (water-cooled), is colourless, fuming liquid, with an ozone-like odour similar to that of some organic compounds that were treated with fluorine. The behavior of the salt is very different from that of a mixture of selenium dioxide and hydrofluoric acid. The liquid quickly attacks glass, and the action with powdered silica is violent:

$$2SeOF_2 + SiO_2 = 2SeO_2 + SiF_4$$

if heated:

$$Si + 2SeOF_2 = SiF_4 + SeO_2 + Se$$

The reaction with phosphorous is:

$$6SeOF_2 + 4P = 4POF_3 + SeO_2 + 5Se$$

An excess of selenium dioxide in selenium oxychloride solution acts like a dehydrating agent as follows:

$$2SeOCL_2 \rightleftharpoons SeO_2 + SeCL_4$$

Phosphorous trichloride produces this vigorous reaction:

$$3SeOCL_2 + 3PCL_3 = SeCL_4 + Se_2CL_2 + 3POCL_3$$

Chromium trioxide or postassium dichromate dissolves in oxydichloride forming a vivid red solution which, when heated, releases fumes of chromyl chloride. At the same time, a small amount of oxydichloride added to lead monoxide or lead dioxide produces noticeable heat and light. The action of selenium oxychloride on lead dioxide is:

$$PbO_2 + 2SeOCL_2 = PbCL_2 + 2SeO_2 + CL_2$$

When dealing with selenium disulphide, one of the important notes is the fact that hydrogen sulphide passed into a cold, diluted solution of selenious acid produces a yellow, pulverulent precipitate containing S : Se = 2:1

$$2H_2S + H_2SeO_3 = 2S + Se + 3H_2O$$

however, if the solution was warm, the reaction is different as sulphuric acid and selenium are formed:

$$2H_2S + 4H_2SeO_3 = 2H_2SO_4 + 4Se + 4H_2O$$

the following step forms deep red precipitate of selenium free from sulphur:

$$2H_2Se + H_2SO_3 = 2Se + S + 3H_2O$$

then the produced sulpur reacts with excess hydrogen selenide as follows:

$$H_2Se + S = H_2S + Se$$

But, gaseous hydrogen selenite passed into sulphurous acid produces orange-red precipitate containg sulphur as follows:

$$6H_2Se + 2H_2SO_3 = 6Se + 2H_2S + 6H_2O$$

The hydrogen sulphide there reacts with the sulphurous acid:

$$2H_2S + H_2SO_3 = 3S + 3H_2O$$

when we reverse the action of sulphuric acid on red amorphous selenium, little sulphoxide is formed:

$$Se + H_2SO_4 \rightleftharpoons SeSO_3 + H_2O$$

The analogy between sulphur sesquioxide, S2O3, and selenium sulphotrioxide, (SeS)O3 is significant. The reason selenium sulphotrioxide is considered within the same window as a selenium sulphite Se(SO3) is the fact that the hydrolysis with water resolves partly into sulphurous acid and partly into what is equivalent to a base:

Se2(SO3)2 + 2H2O = (Se + SeO2) + 2H2O

And partly like silver sulphite into metal and sulphuric acid:

Se(SO3)2 + 2H2O = 2Se + 2H2SO4

Hence, the sulphotrioxide stands in the same relation to the sulphites of strong metals as the corresponding chlorides do to the metal chlorides.

As well, they are not only formed by the direct union of their quasi-metal with sulphur trioxide, but they also can be partially decomposed as follows:

Se2(SO3)2 = 2Se + (SO3)2

Thirdly, it dissolves in fuming sulphuric acid without decomposition, and then it is decomposed like a sulphite by hydrochloric acid as shown below:

Se2(SO3)2 + 2HCL = H2SO3 + CL2Se2(SO3)

And this reacts with hydrochloric acid this way:

Se2(SO3)2 + 2HCL = H2SO3 + CL2Se2(SO3)

And again this reacts with hydrochloric acid:

CL2Se2(SO3) + HCL = Se2CL2 + HCISO3

Laboratory work has shown that when selenium is treated with potassium sulphate in a sealed tube at 130°, potassium monoselenothiosulphate K2SeSO3 is formed. The author confirmed the Rathke experiment by evaporating a solution of selenium in aq. Potassium sulphate. Selenotrithionate was also formed at the same time.

It has been established that selenothiosulphates are isomorphous with thiosuphates. The potassium salt turns brown when heated, forming polysulphide.

The similar behaviour of sulphur dioxide is demonstrated in a neutral solution of silver nitrate, with some selenium precipitates, while ammonial solution reacts this way:

$K_2SSeO_3 + Ag_2O = Ag_2Se + K_2SO_4$

Selenious and sulphurous acids react In aq solutions and selenium is precipitated when the molar propotion $H_2SO_3 = 2:1$ otherwise, a deviation from this ratio, a mixture of monoselenotetrathionic acid $H_2SeS_3O_6$ and diselenotrithionic acid $H_2Se_2SO_6$ is formed. When we have sulphurous acid in excess, se see this:

$3SO_2 + 2H_2O + SeO_2 = H_2SO_4 + H_2S_2SO_6$

however, if selenious acid is in excess:

$2SO_2 + 2SeO_2 + 2H_2O = H_2SO_4 + H_2SSe_2O_6$

meanwhile, selenium acetylacetone reacts quantitatively with sodium or potassium hydrosuphite producing the corresponding alkali monoselenotrithionate as follows:

$(C_5H_6O_2 : Se)_2 + 4NaHSO_3 = 2C_5H_8O_2 + 2Se(NaSO_3)_2$

with sulphurous acid, monoselenotrithionic acid, is formed:

$(C_5H_6O_2 : Se)_2 + 4H_2SO_4 = 2C_5H_8O_2 + 2Se(HSO_3)_2$

What many analysts may have a problem with is the fact that the acid could not be obtained in conc.soln. because, when its aq. soln. is evaporated at normal temperature, red selenium separates when the concentration reaches 50% (or 8N) – while at a higher concentration, the aq. solution deposits selenium and evolves sulphur dioxide even at 0°. If we apply a thallous hydroxide treatment, selenodithionic acid yield thallous selenide and sulphuric acid this way:

$Se(SO_3H)_2 + 2TlOH = Tl_2Se + 2H_2SO_4$

When selenium is dissolved in aq. soln. of potassium sulphite, potassium monoselenotrithionate $K_2S_2SeO_6$ is formed along with some potassium selenothiosulphate. When a soln. of potassium selenothiosuphate and hydrosulphate is evaporated, by dissolving selenium in soln. of potassium sulphite mixed with a little hydrosulphite, or in hydrsulphite alone at 65°, and by adding selenious acid to a soln. of potassium selenothiosulphate mixed with excess potassium sulphite, the warm liquid deposits needle-like crystals of salt on cooling. When they are wahed with cold water, the salt if formed with reasonable purity. This explains what happens:

$2K_2SeSO_3 = K_2SeS_2O_6 + K_2Se$
$2K_2Se + 3SeO_2 = 2K_2SeO_3 + 3Se$
$3SO_2 + SeO_2 + 2H_2O = H_2SO_4 + H_2SeSeO_6$

The salt can also be prepared by action of selenious acid on potassium sulphite:

$3K_2SO_3 + SeO_2 = K_2SeS_2O_6 + K_2SO_4 + K_2O$
and:
$3KSO_3 + 2SeO_2 = K_2SeS_2O_6 + K_2SO_4 + K_2SeO_3$

When selenious acid acts on potassium thiousulphate, both potassium trithionate and selenotrithionate are formed:

$SeO_2 + 4K_2S_2O_3 = K_2S_4O_6 + K_2SeS_4O_6 + 2K_2O$

which is followed by this reaction:

$K_2SeS_4O_6 + 6KOH = 3K_2S_2O_3 + 2K_2SO_3 + 2Se + 3H_2O$

And:
$K_2S_4O_6 + K_2SO_3 = K_2S_3O_6 + K_2SeO_3$

As well as:
$3K_2SO_3 + SeO_2 + H_2O = K_2SeS_2O_6 + K_2SO_4 + 2KOH$

The potassium salt Se(KSO3)2, which is the least soluble of all the alkali monoselenotrithionates, when dry and is heated, decomposes at around 115°, forming selenium, sulphur dioxide, and sulphate:

$K_2SeS_2O_6 = Se + SO_2 + K_2SO_4$

The clear aq. soln. of potassium selenotrithionate decomposes partly to selenium and potassium dithionate, and partly to selenium potassium sulphate, and sulphurous acid. The acid is favoured by boiling the soln., or by evaporating it over conc. sulphuric acid:

$K_2SeS_2O_6 = K_2SO_4 + SO_2 + Se$

With a soln. of barium salt, barium sulphate is precipitated, and on evaporation over water-bath, the salt decomposes:

$BaS_2SeO_6 = BaSO_4 + SO_2 + Se$

Meanwhile, a solution of the barium salt with bromine-water precipitates barium sulphate. In the presence of iodine, potassium selenotrithionate decomposes:

$K_2SeS_2O_6 + 2H_2O + I_2 = K_2SO_4 + H_2SO_4 + Se + 2HI$

With hydrogen sulphide, it reacts as follows:

$K_2SeS_2O_6 + 3H_2S = K_2S_2O_3 + Se + 3S + 3H_2O$

The reaction with copper sulphate is expectedly different:
$K_2SeS_2O_6 + CuSO_4 + 2H_2O = CuSe + K_2SO_4 + 2H_2SO_4$

And, an ammoniacal soln. of a silver salt breacts as follows:

$K_2S_2SeO_6 + Ag_2O + 2NH_3 + H_2O =$
$\qquad K_2SO_4 + Ag_2Se + (NH_4)_2SO_4$

Another note of value to bench analysts relates to common alloy sample analysis. Potassium selenotrithionate behaves with metals as if a selenothiosulphate was first formed for the reason I'll illustrate, and then decomposes either into selenide and sulphate, or, into selenium, sulphur dioxide and sulphate. The action of copper, silver and mercuric salts illustrates this fact:

$2K_2SeS_2O_6 + 3HgCL_2 + 4H_2O =$
$\qquad Hg_3Se_2CL_2 + 4H_2SO_4 + 4KCL$

which is a result of:

$K_2SeS_2O_6 + HgCL_2 = HgSeS_2O_6 + 2KCL$

followed by:

$HgSeS_2O_6 + 2H_2O = HgSe + 2H_2SO_4$

And:

$2HgSe + HgCL_2 = Hg_3Se_2CL_2$

With cadmium sulphgate, lead acetate and ferrous chloride, the reactions are:

$RSeSO_3 + H_2SO_4 = H_2SO_3 + Se + RSO_4$

The reaction between selenium dioxide and sodium thiosulphate has been debated and interpreted by chemists differently. I am confident it is:

$SeO_2 + 4Na_2S_2O_3 = 2NaS_4O_6 + Se + 2Na_2O$

Selenium is precipitated and alkalinity is developed, but in dil. Soln. no selenium is precipitated, and the reaction in not complete due to sodium hydroxide formed neutralizing part of the selenious acid. This makes perfect sense. In th presence of hydrochloric acid the reaction is:

$SeO_2 + 4NaS_2O_3 + 4HCL =$
$\qquad Na_2S_4SeO_6 + Na_2S_4O_6 + 4NaCL + 2H_2O$

No selenium is precipitated, only $Na_2S_4SeO_6$ sodium monoselenopentathionate. Again, this fits logically within the complete picture.

With the aid of alcohol in the cold, you can isolate potassium terathionate and selenopentathionate as follows:

$3SeO_2 + 4K_2S_2O_3 = K_2S_4O_6 + K_2SeS_4O_6 + 2K_2S_2O_3$
followed by:

$2K_2SeS_4O_6 + 6KOH = 3K_2S_2O_3 + 2K_2SO_3 + 2Se + 3H_2O$
by:
$\qquad K_2S_4O_6 + K_2SO_3 = K_2S_3O_6 + K_2S_2O_3$

And by:
$\qquad K_2S_3O_6 + Se = K_2S_3SeO_6$

Potassium tetrathionate and potassium selenotetrathionate are precipitated when a conc. Soln. of selenious acid is mixed with a conc. Soln. of potassium thiosulphate.

And, selenophosphates of the alkali and alkaline metals can be prepared by the action of phosphorus pentaselenide on aq. soln. of the respective metal:

$3R_2Se + P_2Se_5 = 2R_3PSe_4$

I must bring to the attention of the analyst however, that you will not be able to isolate the solid tetraselenophosphate this way, because as soon as it is formed, it is immediately decomposed by water this way:

$R_3PSe_4 + H_2O = H_2Se + R_3POSe_3$

Keeping all the preceding facts and factors in mind will enable you to understand, predict and to succeed.

And now, here are handbook formulas of some selenium compounds that are of importance to the bench analyst:

Cerous dihydrotetraselenite $Ce_2(SeO_3)_3.H_2SeO_3.5H_2O$
Indium hexahydroenneaselenite
$2In_2(SeO_3)_3.3H_2SO_3.12H_2O$
Thallic selenite $Tl_2(SeO_3)_3$
Thallous selenite Tl_2SeO_2
Titanyl dihydroxyselenite $(TiO)_2(HO)_2SeO_3$
Lead oxydiselenitoplumbate $Pb\{PbO(SeO_3)_2\}$

Potassium orthoselenoantimonite K_3SbSe_3 forms in an orange ctystal by evaporating a solution of antimony selenide and potassium selenide in a current of hydrogen.

Sodium orthoselenoantimonite $Na_3SbSe_3.9H_2O$ which crystallizes in yellow needles from a solution of antimony and selenide and sodium selenide.

Potassium selenotetrantimonite $K_2Sb_4Se_7.3H_2O$ a gelatinous brown precipitate.

Tetraselenitohexavanadic acid $H_4V_6O_{17}.4H_2SeO_3$

Ammonium selenitometavanadte $(NH_4)_2O.V_2O_5.2SeO_2$

Sodium decahydrotetraselenitohexavanadate
$Na_2H_2V_6O_{17}.4H_2SeO_3$

Ammonium trihydrotetraselenitohexavanadate
$(NH_4)HV_6O_{17}.4SeO_2.H_2O$

Potassium hemicosihydrodecaslenitohexavanadate
$K_3HV_6O_{17}.10H_2SeO_3.14H_2O$
Potassium heptadecahydroctoselenitohexavanadate
$K_3HV_6O_{17}.8H_2SO_3.4H_2O$

Chromium trioxyenneaselinte $Cr_2O_3.3Cr_2(SeO_3)_3.64H_2O$

Uranyl tetrahydropentaselenite $3UO_3.5SeO_2.7H_2O$

Dimercuriammonium selenate $(NHg_2)_2SeO_4.2H_2O$

Sb

stibnite – trisulfide

stibiconite

cervantite

valentenite

senarmonite

kermesite

sesquioxide

valentenite

senarmonite

kermesite

Antimony

notes

- Antimony is common in soil samples as stibnite {antimony trisulfide), as well as other complex sulfide ores containing lead, copper, mercury and silver.

The two allotropes of antimony are a black amorphous and a yellow covalent formed by oxidation of stibine with oxygen or chloride. The main oxide minerals are stibiconite $Sb_3O_6(OH)x$, cervantite Sb_2O_4 or Sb_2O_3 / Sb_2O_5, valentenite and senarmonite Sb_2O_3, and kermesite Sb_2S_3.

Antimony is oxidized by nitric acid forming a gelatinous precipitate of a hydrated antimony pentoxide. Sulfuric acid forms oxysulfate, while hydrofluoric forms fluorides or fluocomplexes - insoluble compounds.

In the -3 state antimony forms the very unstable compound SbH_3. In the +3 state antimony forms Sb_2O_3 [trioxide or sesquioxide], which at least in one crystal modification, exists as Sb_4O_6 molecules. It is an amphoteric oxide, dissolving in acid to give $Sb(OH)_2^+$ [or SbO^+] ion and dissolves in base to give antimonite anions SbO_2^- or $Sb(OH)_4^-$.

When the sample containing antimonites such as $NaSbO_2$ is acidified, a white precipitate which has the composition $Sb_2O_3.xH_2O$ is formed. It appears that no simple

$Sb(OH)_3$ is formed. The sulfide Sb_2S_3 is orange only when freshly precipitated.

In the +5 state antimony forms pentoxide Sb_2O_5 which is a slightly stronger oxidizing agent than H_3AsO_4.

- Stibine SbH_3, colorless toxic gas with a disagreeable odor. It is produced when metal antimonides are treated with acid, chemical reduction of antimony compounds, and electrolysis of acid or alkaline solutions using a metallic antimony cathode.

$$Zn_3Sb_2 + 6\ H_3O^+ \rightarrow 3\ Zn^{2+} + 2\ SbH_3 + 6\ H_2O$$

$$SbO^{3-}_3 + 9\ H_3O^+ + 3\ Zn \rightarrow SbH_3 + 3\ Zn^{2+} + 12\ H_2O$$

- Alkali metal borohydrides are used to reduce antimony III in acidic aqueous solution to produce stibine. Several metallic antimonides, antimony trioxides, tetraoxides, pentoxides, trifluorides, trichlorides, tribromides, triiodides, trisulfides, pentafluorides, pentachlorides, pentabromides, pentaiodides, pentasulfides play important role in the process of Sb determination by hydrides technique.

Because of the lengthy nature of covering this area, this SOP will only refer to these references:
"Brock University - 1979 paper by Dr. M. Ramzi", "European / Dutch Encyclopedia of chemistry- volume

12", "Cairo / Egypt University research 1970-1980 encyclopedia - v. 23", "EEPA. Ref. v. 4" and BMR v.2 SOP manual by Dr. Paul Gouda of Canada (the author of this book) for details on these compounds.

The analyst needs to be familiar with ways of identifying the presence of these compounds - so that he may approach his sample accordingly.

- The chemistry of heterocyclic antimony compounds and several organoantimony compounds as well as a focus on a few aliphatic primary {$RSbH_2$} and secondary {R_2SbH} / stibines is a major key in the area of antimony recovery.

The reduction of dimethylbromostibine with sodium borohydride produces both methylstibine and dimethystibine. Both are unstable and decompose spontaneously at room temperature.

By contrast, dicyclohexylstibine is stable and is produced as a result of chlorodicyclohexylstibine reduction with lithium aluminum hydride.

- Both phenylstibine and diphenylstibine are common in this type of samples and are easily oxidized.
Diphenylstibine is a strong reducing agent and reacts with acids {in this case - hydride technique, hydrochloric acid} and liberates hydrogen:

$$(C_6H_5)_2SbH + HCL \rightarrow (C_6H_5)_2SbCL_3 + H_2$$

- Antimony III fluoride SbF_3 is a white crystalline or thorhombic solid. It molecule shows a very distorted octahedral arrangement and very soluble in water. Antimony III iodide SbI_3 forms red rhombohedral crystals. Antimony pentafluoride reacts with iodine to form bis(antimony pentafluoride) iodide $Sb_2F_{10}I$. Antimony III sulfide [sesquisulfide] Sb_2S_3 is a black crystalline solid, stibnite.

Antimony pentasulfide appears in the bottom of the test tube as a yellow-orange/redish amorphous solid.

- The hydrolysis of halo-and dihalostibines leads to the formation of compounds of two types $RSbO$ and $(R_2Sb)_2O$.

The aromatic compounds undergo an unusual rearrangements when heated:

$$ArSbO \rightarrow [Ar_2Sb]_2O + Sb_2O_3$$

$$ArSbO \rightarrow Ar_3Sb + Sb_2O_3$$

- A few dialkylstibinic acids exist in soil samples. They are a result of hydrolysis of the corresponding dialkyltrichloroantimony compounds:

$$(CH_3)_2SbCL_3 \rightarrow (CH_3)_2SbO(OH)$$

Aromatic stibonic acids can be produced during sample digestion by the famous diazo reaction:

$$ArSbCL_4 + H_2O \rightarrow ArSbO(OH)_2 + HCL$$

$$ArN_2Cl + SbCl_3 \rightarrow ArSbCl_4 + N_2$$

- When a diazonium salt is present in the sample and is then allowed to react with antimony pentachloride or with an aryltetrachloroantimony compound, the onium salts $[ArN_2][SbCl_6]$ or $[ArN_2][ArSbCl_5]$ are formed. They decompose in organic solvents with formation of diarylantimony trichloride:

$$2[ArN_2][SbCl_6] + 3Fe \rightarrow Ar_2SbCl_3 + 2N_2 + SbCl_3 + 3FeCl_2$$

- Antimony trichloride $SbCl_3$, a colorless crystalline solid soluble in hydrochloric acid and in water when heated, is introduced into the sample as a result of metal chlorination or Sb_2O_3 reaction with HCl conc. It hydrolyzes giving hydrous Sb_2O_3 with excess water but with limited quantities of water a large number of partially hydrolyzed compounds were reported, e.g.. $SbOCl$, Sb_2OCl_4, $Sb_4O_5Cl_2$, $Sb_4O_3(OH)_3Cl_2$, Sb_8O_{11} and Sb_8OCl_{22}.

The hydrolysis precipitation obtained best characterized is tetraantimony dichloride pentoxide $Sb_4O_5Cl_2$.

It is initially precipitated as a thick white solid, changing to well-defined colourless crystals. $SbOCl$ produced changes upon further dilution with water to $Sb_4O_5Cl_2$.

Antimony

... revisited.

Antimony is in the nitrogen group and has an electronegavity of 2.05. As expected by periodic trends, it is more electronegative than tin or bismuth, and less electronegative than tellurium or arsenic. Antimony is stable in air at room temperature, but reacts with oxygen if heated to form antimony trioxide, Sb_2O_3

Four allotropes of antimony are known, a stable metallic form and three metastable forms, explosive, black and yellow. Metallic antimony is a brittle, silver-white shiny metal. When molten antimony is slowly cooled, metallic antimony crystallizes in a trigonal cell, isomorphic with that of the gray allotrope of arsenic. A rare explosive form of antimony can be formed from the electrolysis of antimony(III) trichloride. When scratched with a sharp implement, an exothermic reaction occurs and white fumes are given off as metallic antimony is formed; when rubbed with a pestle in a mortar, a strong detonation occurs.

Black antimony is formed upon rapid cooling of vapor derived from metallic antimony. It has the same crystal structure as red phosphorus and black arsenic, it oxidizes in air and may ignite spontaneously. At 100 °C, it gradually transforms into the stable form. The yellow allotrope of antimony is the most unstable. It has only been generated by oxidation of stibine (SbH_3) at −90 °C. Above this temperature and in ambient light, this metastable allotrope transforms into the more stable black allotrope.

Metallic antimony adopts a layered structure in which layers consist of fused ruffled six-membered rings.

The nearest and next-nearest neighbors form a distorted octahedral complex, with the three atoms in the same double-layer being slightly closer than the three atoms in the next. This relatively close packing leads to a high density of 6.697 g/cm^3, but the weak bonding between the layers leads to the low hardness and brittleness of antimony.

Let's first tour the possible compounds that the analyst may run into during bench preparation of wet chemistry, the sample titration, gravemetric preparation or otherwise any other sample treatment that may include or produce some antimony compounds, some of which might be hazardous.

$Sb(CH_3)_3$ "trimethylstibine", is a colorless pyrophoric and toxic liquid. It is produced by anaerobic bacteria in antimony-rich soils.

Triphenylstibine is the chemical compound with the formula $Sb(C_6H_5)_3$. Abbreviated SbPh3, this colourless solid is often considered the prototypical organoantimony compound. It is used as a ligand in coordination chemistry and as a reagent in organic synthesis.

Like the related molecules triphenylphosphine and triphenylarsine, SbPh3 is pyramidal with a propeller-like arrangement of the phenyl groups. The Sb-C distances average 2.14-2.17 Å and the C-Sb-C angle are 95°.

SbPh3 was first reported in 1886, being prepared from antimony trichlorideby the reaction:

$6\ Na + 3\ C_6H_5Cl + SbCl_3 \rightarrow (C_6H_5)_3Sb + 6\ NaCl$

The modern method employs the Grignard method, using phenylmagnesium bromide and $SbCl_3$.

An antimonite refers to salts of antimony(III), such as NaSb(OH)$_4$ and NaSbO$_2$ (metaantimonite) which can be prepared by reacting alkali with antimony(III) oxide, Sb$_2$O$_3$.

These are formally salts of antimonous acid (antimonious acid), "Sb(OH)$_3$" whose existence in solution is dubious, and attempts to isolate it generally form Sb$_2$O$_3$.xH$_2$O, antimony(III) oxide hydrate, which slowly transforms into Sb$_2$O$_3$.

In geology, the mineral stibnite, Sb$_2$S$_3$, is sometimes called antimonite.

They can be compared to antimonates, which contain antimony in the +5 oxidation state.

Antimony pentachloride is the chemical compound with the formula SbCl$_5$. It is a colourless oil, but typical samples are yellowish due to impurities. Owing to its tendency to hydrolyse to hydrochloric acid, SbCl$_5$ is a highly corrosive substance.

Antimony pentafluoride is the inorganic compound with the formula SbF$_5$. This colourless, viscous liquid is a valuable Lewis acid and a component of the superacid fluoroantimonic acid, the strongest known acid. It is notable for its Lewis acidity and its ability to react with almost all known compounds.

Antimony pentafluoride is prepared by the reaction of antimony pentachloride with anhydrous hydrogen fluoride:

$$SbCl_5 + 5\ HF \rightarrow SbF_5 + 5\ HCl$$

It can also be prepared from antimony trifluoride and fluorine.

In the gas phase, SbF$_5$ adopts a trigonal bipyramidal structure of D3h point group symmetry. The material adopts a more complicated structure in the liquid and solid states.

The liquid contains polymers wherein each Sb is octahedral, the structure being described with the formula [SbF$_4$(μ-F)$_2$]$_n$((μ-F) denotes the fact that fluoride centres bridge two Sb centres).

The crystalline material is a tetramer, meaning that it has the formula [SbF$_4$(μ-F)]$_4$.

SbF$_5$ is a strong Lewis acid, exceptionally so toward sources of F− to give the very stable anion [SbF$_6$]−, called hexafluoroantimonate. [SbF$_6$]− reacts with additional SbF5 to give [Sb$_2$F$_{11}$]−:

SbF$_5$ + [SbF$_6$]− → [Sb$_2$F$_{11}$]−

In the same way that SbF5 enhances the Brønsted acidity of HF, it enhances the oxidizing power of F2. This effect is illustrated by the oxidation of oxygen:

2 SbF$_5$ + F$_2$ + 2 O$_2$ → 2 [O$_2$]+[SbF$_6$]−

Antimony pentafluoride has also been used in the first discovered chemical reaction that produces fluorine gas from fluoride compounds:

4 SbF$_5$ + 2 K$_2$MnF$_6$ → 4 KSbF$_6$ + 2 MnF$_3$ + F$_2$

The driving force for this reaction is the high affinity of SbF5 for F−, which is the same property that recommends the use of SbF5 to generate superacids.

Antimony pentasulfide is an inorganic compound of antimony and sulfur, also known as antimony red. It is a nonstoichiometric compound with a variable composition.

Commercial samples are usually contaminated with sulfur, which may be removed by washing with carbon disulfide in a Soxhlet extractor. It may be used as a red pigment and is one possible precursor to Schlippe's Salt, Na3SbS4, which can be prepared according to the equation:

3 Na$_2$S + Sb$_2$S$_5$ + 9 H$_2$O → 2 Na$_3$SbS$_4$·9H$_2$O

Like many sulfides, this compound liberates hydrogen sulfide upon treatment with strong acids like hydrochloric acid.

6 HCl + Sb$_2$S$_5$ → 2 SbCl$_3$ + 3 H$_2$S + 2 S

Antimony pentoxide (Sb2S5) is a chemical compound of antimony and antimony potassium tartrate oxygen. It always occurs in hydrated form, Sb$_2$O$_5$·nH$_2$O. It contains antimony in the +5 oxidation state.
Antimony potassium tartrate, also known as potassium antimonyl tartrate or emetic tartarhas the formula K$_2$Sb$_2$(C$_4$H$_2$O$_6$)$_2$ is the double salt of potassium and antimony of tartaric acid. The compound has long been known as a powerful emetic. Besides, the compound was used in the treatment of schistosomiasis and leishmania.

Antimony sulfate, Sb$_2$(SO$_4$)$_3$, is a sulfate salt of antimony. This hygroscopic material is formed by reacting antimony or its compounds with hot sulfuric acid. It is used in doping of semiconductors and in the production of explosives and fireworks

Antimony(III) sulfate is a strong oxidizing agent. It is deliquescent, and soluble in acids. It can be prepared by dissolving antimony, antimony trioxide, antimony trisulfide or antimony oxychloride in hot, concentrated sulfuric acid.

$2 \text{ Sb (s)} + 3 \text{ H}_2\text{SO}_4 \rightarrow \text{Sb}_2(\text{SO}_4)_3 + 3 \text{ H}_2$

Antimony telluride is an inorganic compound with the chemical formula Sb_2Te_3

Antimony tetroxide is an inorganic compound with the formula Sb_2O_4. This material, which exists as the mineral cervantite, is white but reversibly yellows upon heating. The material, with empirical formula SbO_2, is called antimony tetroxide to signify the presence of two kinds of Sb centers.

The material forms when Sb_2O_3 is heated in air:

$\text{Sb}_2\text{O}_3 + 0.5 \text{ O}_2 \rightarrow \text{Sb}_2\text{O}_4 \ \Delta H = -187 \text{ kJ/mol}$

At 800 °C, antimony(V) oxide loses oxygen to give the same material:

$\text{Sb}_2\text{O}_5 \rightarrow \text{Sb}_2\text{O}_4 + 0.5 \text{ O}_2 \ \Delta H = -64 \text{ kJ/mol}$

The material is mixed valence, containing both Sb(V) and Sb(III) centers. Two polymorphsare known, one orthorhombic (shown in the infobox) and one monoclinic.

Both forms feature octahedral Sb(V) centers arranged in sheets with distorted Sb(III) centers bound to four oxides.

Antimony tribromide (SbBr_3) is a chemical compoundcontaining antimony in its +3 oxidation state. It may be made by the reaction of antimony with elemental bromine or the reaction of antimony trioxide with hydrobromic acid. It can be added to polymers such as polyethylene as a fire retardant. It is also used in the production of other antimony compounds, in chemical analysis, and in dyeing.

Antimony tribromide has two crystalline forms, both having orthorhombic symmetries. When a warm carbon disulfide solution of SbBr$_3$ is rapidly cooled, it crystallizes into the needle-like α-SbBr$_3$, which then slowly converts to the more stable β form.

Antimony tribromide hydrolyzes in water to form hydrobromic acid and antimony trioxide:

2 SbBr$_3$ + 3 H$_2$O → Sb$_2$O$_3$ + 6 HBr

Antimony trichloride is the chemical compound with the formula SbCl$_3$.

Antimony trifluoride is the inorganic compoundwith the formula SbF$_3$. Sometimes called Swart's reagent, is one of two principal fluorides of antimony, the other being SbF$_5$. It appears as a white solid. As well as some industrial applications, it is used as a reagent in inorganic and organofluorine chemistry.

In solid SbF$_3$, the Sb centres have octahedral molecular geometry and are linked by bridging fluoride ligands. Three Sb-F bonds are short (192 pm) and three are long (261 pm).

Because it is a polymeric, SbF$_3$ is far less volatile than related compounds AsF$_3$ and SbCl$_3$.

SbF$_3$ is prepared by treating antimony trioxide with hydrogen fluoride:

Sb$_2$O$_3$ + 6 HF → 2 SbF$_3$ + 3 H$_2$O
The compound is a mild Lewis acid, hydrolyzing slowly in water. With fluorine, it is oxidized to give antimony pentafluoride.

$SbF_3 + F_2 \rightarrow SbF_5$

Antimony triiodide is the chemical compound with the formula SbI_3. This ruby-red solid is the only characterized "binary" iodide of antimony, i.e. the sole compound isolated with the formula Sb_xI_y. It contains antimony in its +3 oxidation state. Like many iodides of the heavier main group elements, its structure depends on the phase. Gaseous SbI_3 is a molecular, pyramidal species as anticipated by VSEPR theory. In the solid state, however, the Sb center is surrounded by an octahedron of six iodide ligands, three of which are closer and three more distant. For the related compound BiI_3, all six Bi—I distances are equal.

SbI_3 has been used as a dopant in the preparation of thermoelectric materials.
It may be formed by the reaction of antimony with elemental iodine, or the reaction of antimony trioxide with hydroiodic acid.

Antimony trioxide is the inorganic compound with the formula Sb_2O_3. It is the most important commercial compound of antimony. It is found in nature as the minerals valentinite and senarmontite.

Like most polymeric oxides, Sb_2O_3 dissolves in aqueous solutions only with hydrolysis.

Antimony trioxide is mainly produced via the smelting of stibnite ore, which is oxidised to crude Sb_2O_3 using furnaces operating at approximately 850 to 1,000 °C. The transformation is described as follows:
$2\ Sb_2S_3 + 9\ O_2 \rightarrow 2\ Sb_2O_3 + 6\ SO_2$

Crude Sb_2O_3 is purified by sublimation, which allows it to be separated from the more volatile arsenic trioxide. This step is relevant because antimony ores commonly contain significant amounts of arsenic.

Antimony oxide is also obtained via antimony trichloride, which can be obtained from stibnite.

$$2\ Sb_2S_3 + 3\ CaCl_2 + 6\ O_2 \rightarrow 4\ SbCl_3 + 3\ CaSO_4$$

After fractional distillation to separate it from arsenic trichloride, $SbCl_3$ can be hydrolyzed to the oxide:

$$2\ SbCl_3 + 3\ H_2O \rightarrow Sb_2O_3 + 6\ HCl$$

Intermediates in the hydrolysis include the oxychlorides

$SbOCl$ and $Sb_4O_5Cl_2$.

Although impractical for commercial purposes, Sb_2O_3 can be prepared by burning elemental antimony in air:

$$4\ Sb + 3\ O_2 \rightarrow 2\ Sb_2O_3$$

Antimony triselenide is the chemical compound with the formula Sb_2Se_3

Antimony(III) acetate is the antimony salt of acetic acid with the chemical formula of $Sb(CH_3COO)_3$. It has the appearance of a white powder and is used as a catalyst in the production of synthetic fibers. It can be prepared by the reaction of antimony(III) oxide with acetic acid:

$$Sb_2O_3 + 6\ HC_2H_3O_2 \rightarrow 2\ Sb(C_2H_3O_2)_3 + 3H_2O$$

Fluoroantimonic acid ($HSbF_6$) is a mixture of hydrogen fluoride and antimony pentafluoride in various ratios.

The 1:1 combination forms the strongest known superacid, which has been demonstrated to protonate even hydrocarbons to afford carbocations and H_2.

The reaction of hydrogen fluoride (HF) and SbF_5 is exothermic. HF, being a Lewis base, attacks the molecules of SbF_5 to give an adduct. In the fluoroantimonic acid molecule, the anion is coordinated to the hydrogen, although the anion is formally classified as noncoordinating, because it is both a very weak nucleophile and a very weak base.

Gallium indium arsenide antimonide phosphide (GaInAsSbPor GaInPAsSb) is a semiconductor material.

Indium arsenide antimonide phosphide (InAsSbP) is a semiconductor material.
Magic acid (FSO_3H-SbF_5), is a superacid consisting of a mixture, most commonly in a 1:1 molar ratio, of fluorosulfonic acid (HSO_3F) and antimony pentafluoride (SbF_5).

Pentavalent antimonials (also abbreviated pentavalent Sb or SbV) are a group of compounds used for the treatment of leishmaniasis. They are also called pentavalent antimony compounds.

Sodium thioantimoniate, also known as Schlippe's salt, is an inorganic compound with the formula $Na_3SbS_4 \cdot 9H_2O$. This sulfosaltis named after K. F. Schlippe. Sodium thioantimoniate is used to make "quinsulfide antimony," Sb_2S_5. This salt consists of the tetrahedral SbS_4^{3-} anion (rSb-S = 2.33 Å) and sodium cations, which are hydrated. Related salts are known for different cations including ammonium and potassium.

Stibine is the chemical compound with the formula SbH_3. This colourless gas is the principal covalent hydride of antimony and a heavy analogue of ammonia. The molecule is pyramidal with H–Sb–H angles of 91.7° and Sb–H distances of 1.707 Å (170.7 pm). This gas has an offensive smell like hydrogen sulfide (rotten eggs).

Stibophen is an anthelmintic classified as antimony compound and used as treatment of schistosomiasis "intramuscular injection".
Titanium yellow, also nickel antimony titanium yellow, nickel antimony titanium yellow rutile, CI Pigment Yellow 53, or C.I. 77788, is a yellow pigment with the chemical composition of $NiO.Sb_2O_5.20TiO_2$. It is a complex inorganic compound. Its melting point lies above 1000 °C, and has extremely low solubility in water. While it contains antimony and nickel, their bioavailability is very low, so the pigment is relatively safe.

The pigment has crystal lattice of rutile, with 2-5% of titanium ions replaced with nickel(II) and 9-12% of them replaced with antimony(V).

Titanium yellow is manufactured by reacting fine powders of metal oxides, hydroxides, or carbonates in solid state in temperatures between 1000-1200 °C, either in batches or continuously in a pass-through furnace.

Isotopes of antimony

Antimony exists as two stable isotopes, ^{121}Sb with a natural abundance of 57.36% and ^{123}Sb with a natural abundance of 42.64%. It also has 35 radioisotopes, of which the longest-lived is ^{125}Sb with a half-life of 2.75 years. In addition, 29 metastable states have been characterized. The most stable of these is ^{124}Sb with a half-life of 60.20 days, which has an application in some neutron sources

Oxides and hydroxides

Sb_4O_6 "trioxide" is formed when antimony is burnt in air. In the gas phase, this compound exists as Sb_4O_6, but it polymerizes upon condensing. Sb_4O_{10} "pentoxide" can only be formed by oxidation by concentrated nitric acid. Antimony also forms a mixed-valence oxide, antimony tetroxide (Sb_2O_4), which features both Sb(III) and Sb(V). Unlike phosphorus and arsenic, these various oxides are amphoteric, do not form well-defined oxoacids and react with acids to form antimony salts.

Antimonous acid $Sb(OH)_3$ is unknown, but the conjugate base sodium antimonite ($[Na_3SbO_3]_4$) forms upon fusing sodium oxide and Sb_4O_6. Transition metal antimonites are also known. Antimonic acid exists only as the hydrate $HSb(OH)_6$, forming salts containing the antimonate anion Sb(OH)– 6. Dehydrating metal salts containing this anion yields mixed oxides.

Many antimony ores are sulfides, including stibnite (Sb_2S_3), pyrargyrite (Ag_3SbS_3), zinkenite, jamesonite, and boulangertie. Antimony pentasulfide is non-stoichiometric and features antimony in the +3 oxidation state and S-S bonds. Several thioantimonides are known, such as $[Sb_6S_{10}]^{2-}$ and $[Sb_8S_{13}]^{2-}$.

Halides

Antimony forms two series of halides, SbX3 and SbX5. The trihalides SbS3, SbCL13, SbBr3, and SbI3 are all molecular compounds having trigonal pyramidal molecular geometry. The trifluoride SbF₃ is prepared by the reaction of Sb2O3 with HF

$$Sb_2O_3 + 6\ HF \rightarrow 2\ SbF_3 + 3\ H_2O$$

Molten SbF₃ is a weak electrical conductor. The trichloride SbCl₃ is prepared by dissolving Sb₂S₃ in HCL:

$$Sb_2S_3 + 6\ HCl \rightarrow 2\ SbCl_3 + 3\ H_2S$$

Structure of gaseous SbF₅

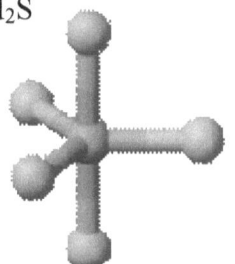

The pentahalides SbF₅ and SbCl₅ have trigonal bipyramidal molecular geometry in the gas phase, but in the liquid phase, SbF₅ is polymeric, whereas SbCl₅ is monomeric.

SbF₅ is a powerful Lewis acid used to make the superacid fluoroantimonic acid "HSbF₆".

Oxyhalides are more common for antimony than arsenic and phosphorus. Antimony trioxide dissolves in concentrated acid to form oxoantimonyl compounds such as SbOCl and (SbO)₂SO₄.

Antimonides, hydrides, and organoantimony compounds

Compounds in this class generally are described as derivatives of Sb^{3-}. Antimony forms antimonides with metals, such as indium antimonide (InSb) and silver antimonide (Ag_3Sb). The alkali metal and zinc antimonides, such as Na_3Sb and Zn_3Sb_2, are more reactive. Treating these antimonides with acid produces the unstable gas stibine SbH_3

$$Sb^{3-} + 3\ H^+ \rightarrow SbH_3$$

Stibine can also be produced by treating Sb^{3+} salts with hydride reagents such as sodium borohydride. Stibine decomposes spontaneously at room temperature. Because stibine has a positive heat of formation, it is thermodynamically unstable and thus antimony does not react with hydrogen directly.

Organoantimony compounds are typically prepared by alkylation of antimony halides with Grignard reagents. A large variety of compounds are known with both Sb(III) and Sb(V) centers, including mixed chloro-organic derivatives, anions, and cations. Examples include $Sb(C_6H_5)_3$ "triphenylstibine", $Sb_2(C_6H_5)_4$ (with an Sb-Sb bond), and cyclic $[Sb(C_6H_5)]_n$. Pentacoordinated organoantimony compounds are common, examples being $Sb(C_6H_5)_5$ and several related halides.

Ancient Egyptian hieroglyphs dating from about 3000 years ago (1890 BC) at the top of this wall painting say: "The arrival, bringing stibium." The Egyptian word for stibium was mestchem-t.

Antimony III sulfide, Sb_2S_3, was recognized in predynastic Egypt as an eye cosmetic (kohl) as early as about 3100BC, when the cosmetic palette was invented. Copper objects plated with antimony dating between 2500 BC and 2200 BC Egypt.

The Egyptians called antimony *mśdmt*; in hieroglyphs, the vowels are uncertain, but there is an Arabic tradition that the word is *mesdemet*. The Greek word, στίμμι *stimmi*, is driven from Arabic or Egyptian *sdm*

Antimony can be isolated from the crude antimony sulfide by a reduction with scrap iron:

$$Sb_2S_3 + 3\ Fe \rightarrow 2\ Sb + 3\ FeS$$

The sulfide is converted to an oxide and advantage is often taken of the volatility of antimony(III) oxide, which is recovered from roasting. This material is often used directly for the main applications, impurities being arsenic and sulfide. Isolating antimony from its oxide is performed by a carbothermal reduction:

$$2\ Sb_2O_3 + 3\ C \rightarrow 4\ Sb + 3\ CO_2$$

Antimony and many of its compounds are toxic, and the effects of antimony poisoning are similar to arsenic poisoning. The toxicity of antimony is by far lower than that of arsenic; this might be caused by the significant differences of uptake, metabolism and excretion between arsenic and antimony. The uptake of antimony(III) or antimony(V) in the gastrointestinal tract is at most 20%. Antimony(V) is not quantitatively reduced to antimony(III) in the cell. Since methylation of antimony does not occur, the excretion of antimony(V) in urine is the main way of elimination. Reported cases of intoxication by antimony equivalent to 90 mg antimony potassium tartrate dissolved from enamel has been reported to show only short term effects. An intoxication with 6 g of antimony potassium tartrate was reported to result in death after 3 days.

Inhalation of antimony dust is harmful, and in certain cases may be fatal; in small doses, antimony causes headaches, depression and dizziness. Larger doses such as prolonged skin contact may cause dermatitis, or damage the kidneys and the liver, causing violent and frequent vomiting leading to death in a few days.

Antimony is incompatible with strong oxidizing agents: strong acids, halogen acids, chlorine or fluorine. Therefore, it is kept away from heat.

Antimony leaches from polyethylene terephthalate (PET) bottles into liquids. While levels observed for bottled water are below drinking water guidelines, fruit juice concentrates (for which no guidelines are established) produced in the UK were found to contain up to 44.7 μg/L of antimony, well above the EU limits for tap water of 5 μg/L.

Atomic Structure

Number of Energy Levels: 5

First Energy Level: 2
Second Energy Level: 8
Third Energy Level: 18
Fourth Energy Level: 18
Fifth Energy Level: 5

Atomic Number: 51
Atomic Mass: 121.76 amu

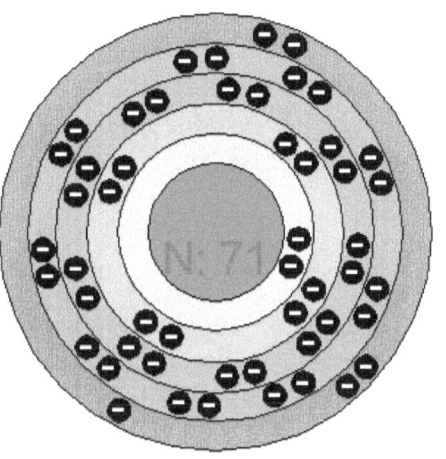

Melting Point: 630.0 °C (903.15 K, 1166.0 °F)
Boiling Point: 1750.0 °C (2023.15 K, 3182.0 °F)

Number of Protons/Electrons: 51
Number of Neutrons: 71

Classification: Metalloid
Crystal Structure: Rhombohedral
Density @ 293 K: 6.684 g/cm^3

Antimonide minerals:

Allargentum is an antimonide mineral, superclass of sulfides and sulfosalts, with formula written as $Ag_{1-x}Sb_x$, where x = 0.09–0.16.

An antimonide mineral contains antimonide as its main anion.

Aurostibite is an isometric gold antimonide mineral which is a member of the pyrite group.

Breithauptite is a nickel antimonide mineral with the simple formula NiSb. Breithauptite is a metallic opaque copper-red mineral crystallizing in the hexagonal - dihexagonal dipyramidal crystal system.

The silver antimonide mineral dyscrasitehas the chemical formula Ag_3Sb

Stibiopalladinite is a mineral containing the chemical elements palladium and antimony. Its chemical formula is Pd_5Sb_2.

Ullmannite is a nickel antimont sulfide mineral with formula: NiSbS.

Antimonides:

Aluminium antimonide (AlSb) is a semiconductor of the group III-V family containing aluminium and antimony.

Antimonides (also known as stibnides) are compounds of antimony with more electropositive elements. The antimonide ion is Sb^{3-}.
Many of them are flammable or decomposed by oxygen when heated since the antimonide ion is a reducing agent. An antimonide mineral is a mineral that contains antimonide for its main anion. The antimonides are structurally similar to the sulfides and are grouped with them in both the Dana and Strunz mineral classification systems.

Aurostibiteis an isometric gold antimonide mineral which is a member of the pyrite group.

Gallium antimonide (GaSb) is a semiconducting compound of gallium and antimony of the III-V family. It has a lattice constant of about 0.61 nm.

Indium antimonide (InSb) is a crystalline compound made from the elements indium(In) and antimony(Sb). It is a narrow-gap semiconductor material from the III-V group used in infrared detectors, including thermal imaging cameras, FLIR systems, infrared homing missile guidance systems, and in infrared astronomy Yttrium(III) antimonide (YSb) is an inorganic chemical compound.

Zinc antimonide (ZnSb), (Zn_3Sb_2), (Zn_4Sb_3) is an inorganic chemical compound. Like indium antimonide, aluminium antimonide, and gallium antimonide, it is a semiconducting intermetallic compound. It is used in transistors, infrared detectors and thermal imagers, as well as magnetoresistive devices.

Other antimony notes:

Antimony forms the volatile trichloride, and a nonvolatile antimony compound when heated with chlorosulphonic acid - with sulphuryl chloride, the reaction is:

$$2Sb + 3SO_2CL_2 = 2SbCL_3 + 3SO_2$$

with thionyl chloride:

$$6Sb + 6SOCL_2 = 4SbCL_3 + Sb_2S_3 + 3SO_2$$

and with pyrosulphuryl chloride:

$$2Sb + 3S_2O_5CL_2 = 2SbCL_3 + 3SO_2 + 3SO_3$$

when antimonic acid is treated with insufficient quantity of hydrogen sulphide, the antimonic acid is reduced to antimonious acid:

$H_3SbO_4 + H_2S + Aq. = H_3SbO_3 + H_2O + S + Aq.$

The reason is the fact that hydrogen sulphide does not reduce antimonic acid, and that its sulphur is likely to combine with the antimony as sulphantimonic acid. As well, even in the presence of a strong reducing agent such as sulphur dioxide, only about 70% of the antimonic acid is reduced to antimonious acid. It is possible that when an excess of hydrogen sulphide is present along with free mineral acid, sulphantimonic acid is first formed:

$3H_3SbO_4 + 8H_2S + Aq. = 2H_3SbS_4 + 8H_2O + Aq.$

then, is followed by the formation of antimony pentasulphate:

$2H_3SbS_4 + nHCL + Aq. = Sb_2S_5 + 3H_2S + nHCL + Aq$

However, I think it is more likely that when the antimonic acid is in excess, sulphantimonic acid is first formed, but, in contact with the excess free antimonic acid, it forms trioxysulphantimonic acid this way:

$H_3SbS_4 + 3H_3SbO_4 + Aq. = 4H_3SbO_3S + Aq.$

This however must be a very unstable compound and as such it decomposes at once into antimonious acid and sulphur:

$nH_3SbO_3S + mHCL + Aq. = nH_3SbO_3 + nS + mHCL + Aq.$

A soln. of an alkali sulphide with antimonious salts gives a precipitate of orange-red antimony trisulphide - and orange antimony pentasulphide fron soln. of antimonic salts. All salts of antimony, when warmed with sodium thiosulphate, give a precipitate of the sulphide:

$2SbCL_3 + 3NaS_2O_3 + 3H_2O = Sb_2S_3 + 3Na_2SO_4 + 6HCL$

Unlike soln. of antimonious salts, antimony salt soln. are reduced by hydriodic acid or alkali iodides in acid soln.

Sodium hypochloride soln. slowly oxidizes stibine. While iodic acid also oxidizes it.

Phosphorus triiodide reacts with stibine producing diiodide – and phosphorus diiodide reacts:

$3P_2I_4 + 4SbH_3 = 4PH_3 + 4SbI_3 + 2P$

then the antimony triiodide reacts:

$SbH_3 + SbI_3 = 2Sb + 3HI$

Antimony tetroxide liberates iodine from a mixture of potassium iodide and hydrochloric acid.:

$Sb_2O_4 + 6HCL + 2HI = 2SbCL_3 + 4H_2O + I_2$

The conversion of $Sb_4O_3(OH)_3CL_3$ into $SbOCL$ and $Sb_4O_5CL_2$ is as follows:

$Sb_4O_3(OH)_3CL_3 + HCL = 4SbOCL + 2H_2O$
And

$Sb_4O_3(OH)_3CL_3 = Sb_4O_5CL_2 + H_2O\ HCL$

The reaction representing the production of metallic antimony or regulus from crude antimony, or from rich ores, with over 60% Sb, which repeated at bench sample treatment, can be affected by the precipitation process, or iron reduction process:

$Sb_2S_3 + 3Fe = 2Sb + 3FeS$

In the process of direct smelting of stinite, no iron, when was used in the reduction, the molten sulphide dissolves the trioxide formed:

$2Sb_2S_3 + 9O_2 = Sb_4O_6 + 6SO_2$

which also involves the following reaction:

$Sb_4O_6 + Sb_2S_3 = 6Sb + 3SO_2$

When a layer of antimony as fine powder in a flat-bottomed flask is added to a layer of dry ether and added to 5 parts of bromine, the mixture is then heated with a reflux condenser for 3 hours, crystals of antimony ethyl oxypentabromide $SbBr_5(C_2H_5)_2O$ are formed. When these crystals are allowed to stand over sulphuric acid, they lose ether above 55° forming antimony pentabromide $SbBr_5$ which loses bromine at 80°.

The tribromide can be obtained by heating a mixture of antimony sulphate and dry potassium bromide as follows:

$Sb_2(SO_4)_3 + 6KBr = 3K_2SO_4 + 2SbBr_3$

In dealing with many organic samples, the analyst's mental or written notes must include that antimony trisuphide is quickly reduced - with heat - by carbon to antimony, forming carbon disulphide. The following reaction in the presence of potassium nitrate is common:

$Sb_2S_3 + 8KNO_3 + 6C =$
$\qquad 2Sb + 3K_2SO_4 + K_2CO_3 + 4CO_2 + CO + 4N_2$

When looking at the pattern of many metal oxides when mixed with carbon (reduction of the oxide to metal) – we see that mercuric oxide transforms the trisulphide into antimony pentoxide. When an excess of alkali hydroxide or carbonate is heated with antimony trisulphide, antimony pentoxide and an alkali sulphide are formed – the antimony pentoxide forms either an oxysulphantimonate or an antimonite with the undecomposed alkali and trisulphide, while the alkali sulphide forms a sulphantimonite.

The reaction with an excess of carbonate is as follows:

$5Sb_2S_3 + 7K_2CO_3 = 7CO_2 + 3K_4Sb_2S_5 + K_2Sb_4O_7$

and with an excess of trisulphide:

$4Sb_2S_3 + 3K_2CO_3 = 6KSbS_2 + Sb_2O_3 + 3CO_2$

some antimony may be reduced this way:

$2K_2Sb_4O_7 = 2Sb + 3Sb_2O_4 + 2K_2O$

or, another possibility is:

$5K_4Sb_2S_5 = 4Sb + K_3SbS_4 + K_2S$

And, when digesting a sample in a test tube or a flask, keep in mind that crystalline trisulphide, when boiled with alkali-lye, the alkali sulphide and antimony oxide first produced are converted to antimonites and sulphantimonites.

The pyrosulphantimonite can dissolve more trisulphide:

$K_2Sb_2S_5 + Sb_2S_3 = 4KSbS_2$

However, when the solution is cooled, the pyrosulphantimonite is formed, and the trisulphide is deposited as "mineral kermes."

Treating an excess of antimony trisulphate with a boiling soln. of ammonium hydrosulphide, adding alcohol, you precipitate a pale yellow mass of rhombohedral crystals. If alcohol is added to a mol of antimony trisulphide and conc. Soln. of 3 mols of ammonium sulphide, a white crystalline precipitate of ammonium orthosulphoantimonite $(NH_4)_3SbS_3$ is formed. If you crystallize the solution without the addition of alcohol, ammonium metasulphoantimonite $NH_4SbS_2.2H_2O$ is formed.

You can, and I did, obtain the same salt by treating pulverulent stibnite with a freshly prepared solution of ammonium sulphide. you'd obtain yellow crystals which in about 10 hours are drained from the host liquor, washed with water, then with alcohol and ether, and dried between bibulous paper. The yellow, four-sided needles are fluorescent and they become brownish-red when exposed to air. They are insoluble in water and at 105° after losing weight, they become ammonium metasulphotetrantimony $(NH_4)_2Sb_4S_7$. Heating ammonium orthousulphoantimonate and antimony pentasulphide in a sealed test tube also produced the same salt:

$$Sb_2S_5 + 2(NH_4)_3SbS_4 = 2(NH_4)_2S_3 + (NH_4)_2Sb_4S_7$$

When heating sodium thiosulphate with a conc. Soln. of sodium antimonite, sodium pyroantimonate, sulphantimonate, and sulphite are formed. If action with potassium salts takes place in the cold, antimony trisulphide is precipitated and by evaporating the hosting liquor, we get crystals of what is probably the hydrodioxydisulphoantimonate. I have to say "probably" because small details and other factors including the chemists' bench attitude play a role here, and because this same salt is produced by the action of potassium hydroxide and sulphur on antimony trioxide as follows:

$$Sb_2O_3 + 5KOH + 2S = K_2HSbO_2S_2 + KH_2SbO_4 + 2KOH$$

I have achieved the same result utilizing the action of potassium hydroxide and polysulphide on antimony trioxide this way:

$$3Sb_2O_3 + 18KOH + 22S_4 =$$
$$3K_2HSbO_2S_2 + 3K_2SbO_4 + 2K_2S + 9KOH$$

Related analytical notes:

With the preceding facts in mind, let's now go back to the AAS / ICP bench. You have prepared the sample correctly and safely, and have completed the sample digestion and dilution. The AAS / ICP is optimized and calibration graph provides .995 to .999 C.C.

You have your unknown or target sample, along with:
- Blank / analytical zero; including a digested blank.
- 4 standards, e.g. 1, 3, 5, 10 ppb
- quality control sample-Q1: a known value, previously certified sample of the same matrix, e.g. soil, sludge, lake water, seafood, pharmaceutical drug ... etc.
- Q2: a certified standard, e.g. BDH
- Q3: a duplicate sample spiked with a known standard edition, e.g. additional 1 ppb
- Q4 & 5 additional certified standards digested in the same fashion as samples: A processed matrix-match blank, and 3 ppb MM std.

With all the preceding facts in mind, the following notes will now come into effect.

- The sample and the HCL are allowed to merge first before the sodium borohydride enters the stream to meet the sample in the reaction tube.

- Fresh borohydride solution must be prepared on the analysis day to ensure its stability. Sodium hydroxide is added to stabilize the solution. Allow solution to reach room temperature prior to application.

When determining As / Se in soils samples containing high concentration of metals such as Co, Fe, Ni, fewer interference have been observed when 0.3 % w/v NaBH4 solution is used as reductant { rather than 0.7 - 1.0 % }.

This was achieved at the cost of reducing sensitivity by lowering the concentration of the borohydride solution. The analyst need to consider his option according to the sample nature.

- Varian study was based on As^{III} in 7 M HCL , Sb^{III} in 7 M HCL & Se^{IV} in 7 M HCL. The MOE study involved the use of 1 - 10 M HCL. The experiments indicated the following:

With 10 M HCL acid channel / As at 193.7 nm wavelength:

Solutions were prepared in 1 M hydrochloric acid and the analyte was reduced to As^{III} by the action of potassium iodide at a concentration of 1% w/v.

Reduction required about 50 minutes at room temperature or about 4 minutes at 70 ^0C. When the analyte was retained as As^V by omitting the reduction step, the analytical response was only 20% of that observed as As^{III}.
At 10% KI concentration and samples prepared in 7 M hydrochloric As^V responded the same as As^{III}

- With 10 M HCL / Sb at 217.6 nm wavelength:
Sb^{III} solutions were prepared in 1 M HCL and gave good recovery up to 40 ug/L. With a 10% KI solution in 7 M

HCL and a second set in 10 M HCL, good recovery was secured.

- Se^{VI} was reduced to Se^{IV} by warming with concentrated HCL. In the IV oxidation state in 1 M HCL, Se showed good response. In the VI oxidation state no response was detected.

- AAS optimization is imperative. Allow the hollow cathode lamp to warm up for some 30 minute prior to calibration - depending on lamp condition. Ensure that the lamp is producing sufficient voltage impressed across the electrodes, i.e. its ability to ionize sufficient Ar atoms enough to bombard the cathode and thus produce an effective electromagnetic radiation [energy beam]. At times, this may necessitate alteration to AAS setting to compensate for such problem.

Since the intensity of light source is proportional to the square of lamp current, the AAS settings for the specific sample must consider the sample matrix and any false positive of signal suppression that may result. This also applies to baseline noise level - lamp current relation which must be monitored for the possible absorbance of other elements present in the sample that fell within the width of the monitored element's absorbance line. The lamp also can be affected by impurities in the cathode itself or infected from the "W" anode.

It should also be noted that an increase in the cathode lamp current results in an increase in the kinetic energy of the ionized fill gas "Ar" causing more atoms to be sputtered. As the population of the sputtered atoms increases, the residual unexcited atoms cool and a cloud of neutral atoms in front of the cathode is formed. These neutral atoms absorb some of the lamp light which results in an attenuation of the resonance radiation resulting in a classic case of self reversal or self absorption.

- Adjust the wavelength at one slit bandwidth narrower than operation setting then open the bandwidth wider. Spectral bandwidth must be adjusted to suite the operation needs. A large width will generate a good signal-to-noise ratio, however the resonance line may not be isolated from other lines and as a result the analysis curve may not be as linear. At the same time, the good resolution of a too narrow spectral band width will not compensate for the poor signal-to-noise ratio due to the reduction of light. The attached method summary provides a good guideline; but , at times - with certain samples - it must altered to suit.

- Allow the absorbance cell to stablize with the flame for sufficient time. This is followed by conditioning of the cell Auto-zero , then read the blank [fresh - undigested 5% HCL] to monitor the baseline. Run a high [fresh-undigested] standard (e.g. 20, 50 or max. 100 ug/L) and secure an obvious signal. Run the blank again to ensure a representative response.
Repeat the blank & standard runs 2 or 3 times to confirm the conditioning of the cell. This is followed by auto-zero with digested - analytical blank.

- Ensure that the computer interface corresponding to hydride data is connected. Also ensure the connection of the hydride argon line. Confirm the stability of the argon, acetylene and compressed air pressure as stated herein . Any change in the acetylene or air pressure will affect the flame type and temperature. Recheck the flow rate of the DD water to the nebulizer borohydride suction, hydrochloric suction & sample suction. Adjust the pump and the tubes to achieve the rates stated earlier in this reference.

Sequence of instrumental (AAS) analysis

1]- Auto-zero, 5% HCL solution
2]- blank, 5% HCL solution
3]- cell conditioning with repeated runs of blank and standard

4]- auto zero with analytical - digested / matrix match blank
5]- calibration : digested blank , digested standards : 1, 3, 5 ug/l.
6]- blank , digested / matrix - match {Q1}

7]- digested std "3 ug/l" {Q2}
8]- digested matrix-match QC to confirm the acceptance of the standard solution, the calibration and the digestion . {certified / digested QC soil}.

9]- first sample "D"
10]- samples, "average of a 10 samples set"
11]- replicant of "D" sample , Q4

12]- Spiked "D" sample, Q3
13]- repeat of blank - Q1

14]- repeat of Q2
15]- programmed auto-zero and auto-calib.
16]- monitor low and high point on calib.
    curve {e.g. std. 1 & 5 ppb}

Interference:

- High concentrations of chromium, nickel, copper, mercury, molybdenum, silver, cobalt or tin cause depression of the signal.

- Traces of nitric results in analytical interference.

- Confirm that the reagents used are contamination free {via blank}.

Optimization:

Part -2:   [Operation]

1] after lamp warm-up [at the specified lamp current], fine-adjust the wavelength {at 1 setting lower than operation setting as indicated. At times ,at 2 settings lower} and at single beam.

2] adjust the cathode lamp horizontal & vertical to maximize signal.

3] open the slit to operational setting

4] switch to double beam

5] adjust the gain voltage to obtain the ideal setting

6] switch back to single beam and place the quartz cell in the AAS optical path and monitor and correct /

maximize the signal. Visually monitoring the beam for initial alignment -e.g. with aid of card- can be helpful.

7] Switch back to double beam if you have not done so already

8] background correction: no programmed method in this case.

9] Flame mixture adjusted as stated in this literature. Flame / cell are allowed to stabilize .

10] confirm the optimization based on response with final touches.

Routine maintenance:

1- The quartz cell must be cleaned regularly and soaked between runs. Se cell must be kept separately from As/Sb cell to avoid any KI contamination.
2- The pump VG-76 unit must be checked and maintained regularly. Calibration of suction rate is imperative.

3- Two separate sets of tubes and the glass separators for "Se" & "As/Sb" must be cleaned and replaced as often as needed.

4- General & basic maintenance of the AAS at large.

Software issues

The EPL AAS computer system is programmed to perform 4 methods; two for normal operational level "0-5 ug/L in solution" and two methods for a high level of up to 10 ppb as follows. The following are the actual methods created and utilized by chemist-author.

In interpreting the method name:
"PG" stands for Paul Gouda; while "HG" stands for hydride generation and "GA" stands for ghraphite furnace, both are Atomic absorption methods.

These methods are still in use by analytical laboratories utilizing the atomic absorption technique. It is understood of course, that the likely use of a different AAS model will necessitate altering the following setup accordingly.

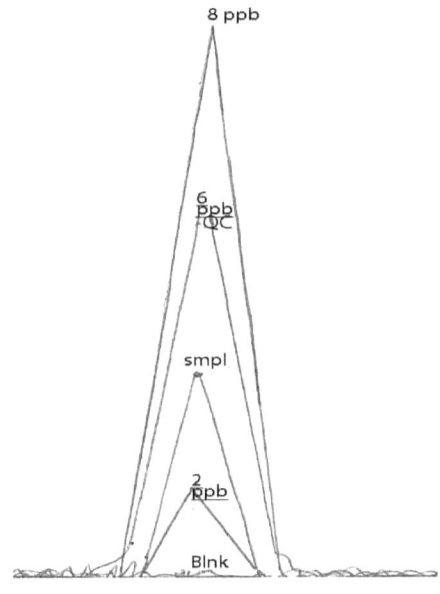

Method I:

Method name: **PGASSBHG**
(and method report format)

Paul Gouda / As&Sb / hydride generation

Elements: As Sb

Atomizer: Flame
Matrix: ULTRAPURE WATER

METHOD INFORMATION **

Default Setup:

   Number of Repeats: 1
   Flush Time (sec)   : 45.0
   Auto-Increment Sample Names? No

Auto-print Calibration Curves? : Yes
Analysis Graphics Display absorbance
Auto-print Analysis Graphics : Yes
Auto-save Analysis Graphics   : No

Default File Names:

Analysis Data File : RESULTS   Sample Limits
Table   : LCTAB
Autosampler Table : ELEMENT   Blank (AZ) Limits
Table   : BLANKLCT
          QC Check Table   : QCTAB
                                        Recovery
Check Table   : RQCTAB

OUTPUT INFORMATION**

Output Mode :   Concentration
Override Print Limits?   Yes
Override Significant Figures?   No
Limits Table : LCTAB
Check?   No
Correction Factor: 1

Auto-print data?   Yes
Condensed report format?   Yes
Auto-store data?   Yes
Store individual repeats?   No

Report to:
Screen   Avgs, Stats, Units
Printer   Avgs, Stats, Units

Mode : Double Beam Atomic Absorption
Integ Type   : Automatic   Integ Time : 3.0 Sec.
     Delay Time: 0.0 Sec.
Comment   : ARSENIC/ANTIMONY (HYDRIDE GENERATION) designed by Chemist Paul Gouda, Ontario, Canada.
Flame Information:
Flame Type   : Air/Acetylene
Oxidant Flow (SCFH)   : 5.0 - 2.5

Element   : *As*

Wavelength   : 193.70
Bandpass   : 2.0
High Voltage   : 620
BKG Method   : None

Lamp Current	: 6.0
Signif Figs	: 4
Print Units	: ug/L
Print Limits Low	: .2
Print Limits High	: 0

Stdzn Method: Multipoint Stdn.
Stdzn Method: Standard Additions

Std Names	Std Conc	Abs.
Addn Name	Std Addn	
A/Z#1: BlankSTD	= 0.00	= 0.0000
#1: Spike1	= 10.00	
#2: ASSESB-1	= 1.00	= 0.0582
#2: Spike2	= 20.00	
A/C#3: ASSESB-3	= 3.00	= 0.1519
#3: Spike3	= 30.00	
#4: ASSESB-5	= 5.00	= 0.2429

Element	: *Sb*
Wavelength	: 217.60
Bandpass	: 0.50
High Voltage	: 530
BKG Method	: None
Lamp Current	: 10.0
Signif Figs	: 4
Print Units	: ug/L
Print Limits Low	: .2
Print Limits High	: 0

Stdzn Method: Multipoint Stdn.
Stdzn Method: Standard Additions

Std Names	Std Conc	Abs.	Addn Name	Std Addn
A/Z#1: BlankSTD	= 0.00	= 0.0000	#1: Spike1	= 10.00
#2: ASSESB-1	= 1.00	= 0.0504	#2: Spide2	= 20.00
A/C#3: ASSESB-3	= 3.00	= 0.1414	#3: Spike3	= 30.00
#4: ASSESB-5	= 5.00	= 0.2352		

Method II: **PGSEHG**

Method Report
Paul Gouda / Se / hydride generation

Atomizer: Flame
Matrix: ULTRAPURE WATER

METHOD INFORMATION **

Default Setup:

    Number of Repeats   : 1
    Flush Time (sec)   : 40.0
Auto-Increment Sample Names? No

Auto-print Calibration Curves? : Yes
Analysis Graphics Display   : Absorbances
Auto-print Analysis Graphics? : Yes
Auto-save Analysis Graphics? : No

Default File Names:

Analysis Data File: RESULTS
Sample Limits Table: LCTAB
Autosampler Table: HYDRIDE
Blank (AZ) Limits Table   BLANKLCT

QC Check Table   : QCTAB
Recovery Check Table   : RQCTAB
OUTPUT INFORMATION **

Output Mode:	Concentration
Override Print Limits?	Yes
Override Signif Figs?	No

Limits Table: LCTAB   Check?   No
Correction Factor: 1

Auto-print data?   Yes
Condensed report format?   Yes
Auto-store data?   Yes
Store individual repeats?   No

Report to:
    Screen   Avgs, Stats, Units
    Printer   Avgs, Stats, Units

Method: **PGSEHG**
part -2-

Paul Gouda / Se / hydride generation

Element   : Se

Mode   : Double Beam Absorption

Integ Type: Automatic   Integ Time: 3.0 Sec.
Delay Time: 0.0 Sec.

Comment   : SELENIUM BY HYDRIDE GENERATION.

Flame Information:

Flame Type   : Air / Acetylene
Oxidant Flow (SCFH)   : 5.0
Oxidant Flow (SCFH)   : 2.5
Flame Type   : Air/Acetylene

Element   : Se
Element Name   : Se
Wavelength   : 196.00
Bandpass   : 2.0
High Voltage   : 700
BKG   : None
Lamp Current   : 4.0
Signif Figs   : 4
Print Units   : ug/L
Print Limits Low   : 0
Print Limits High : 0

Stdzn Method: Multipoint Stdn.
Stdzn Method: Standard Additions
Std Names   Std Conc
Abs.
Addn Name   Std Addn
A/Z#1: BlankSTD   = 0.00   =

0.0000   #1: Spike1   = 10.00
#2: ASSESB-1   = 1.00   = 0.0274
#2: Spike2   = 20.00

A/C#3: ASSESB-3   = 3.00   = 0.0851
    #3: Spike3   = 30.00

#4: ASSESB-5   = 5.00   = 0.1401

Method III: **PGASSBHGA**

Paul Gouda / As / Sb Hydride generation, Graphite furnace, Atomic Absorption

Method Report

As / Sb / Graphite "furnace" **A**tomic Absorption

Atomizer: Flame
Matrix: ULTRAPURE WATER

METHOD INFORMATION **

Default Setup:

    Number of Repeats : 1
    Flush Time (sec)   : 45.0
    Auto-Increment Sample Names? No

    Auto-print Calibration Curves? : Yes
    Analysis Graphics Display   : Absorbances
    Auto-print Analysis Graphics? : Yes
    Auto-save Analysis Graphics?   : No

Default File Names:
    Analysis Data File : RESULTS   Sample Limits Table   : LCTAB
    Autosampler Table : HYDRIDE   Blank (AZ) Limits Table : BLANKLCT
QC Check Table   : QCTAB
Recovery Check Table   : RQCTAB

OUTPUT INFORMATION:
Output Mode: Concentration
Override Print Limits?   No

Override Sidnif Figs   No
Limits Table: LCTAB   Check? No
Correction Factor: 1

Auto-print data?   Yes
Condensed report format?   Yes
Auto-store data?   Yes
Store individual repeats?   No

Report to:
  Screen   Avgs, Stats, Units
  Printer   Avgs, Stats, Units

Method IV: PGASSBHGA   part-2
Paul Gouda / As / Sb / Hydride Generation.

Elements   : As   Sb
Mode   : Double Beam Absorption
Integ Type   : Automatic   Integ Time : 3.0 Sec.
Delay Time : 0.0 Sec.

Comment   : ARSENIC/ ANTIMONY BY
Graphite furnace - Atomic Absorption

Flame Information:
Flame Type   : Air/Acetylene
Oxidant Flow (SCFH)   : 10.0
Oxidant Flow (SCFH)   : 4.5
Flame Type   : Air/Acetylene

Element   *: As*
Wavelength   : 193.70
Bandpass   : 2.0
High Voltage   : 620
BKG Method   : None
Lamp Current   : 6.0
Signig Figs   : 4
Print Units   : ug/L
Print Limits Low   : 0
Print Limits High   : 0

Stdzn Method: Multipoint Stdn.
Stdzn Method: Standard Additons

Std Names   Std Conc   Abs.
Addn Name   Std Addn
A/Z#1: BlankSTD   = 0.00   = 0.0000
    #1: ADD1   = 10.00

#2: ASSESB-1   = 1.00   = 0.0341
#2: ADD2   = 20.00

A/C#3: ASSESB-5   = 5.00   = 0.1610
#3: ADD3   = 30.00

#4: ASSESB-10   = 10.00   = 0.2621

Element   : Sb

Wavelength   : 217.60
Bandpass   : 0.50
High Voltage   : 620
BKG Method   : None
Lamp Current   : 10.0
Signif Figs   : 4
Print Units   : ug/L
Print Limits Low : 0
Print Limits High : 0

Stdzn Method: Multipoint Stdn.
Stdzn Method: Standard Additions

Std Names   Std Conc   Abs.
Addn Name   Std Addn
A/Z#1: BlankSTD   = 0.00   = 0.0000

#1: ADD1   = !0.00
#2: ASSESB-1   = 1.00   = 0.0259
#2: ADD2   = 20.00

A/Z#3: ASSESB-5   = 5.00   = 0.1328
#3: ADD3   = 30.00

#4: ASSESB-10 = 10.00   = 0.3393

Method V):   **PGSEHGA**
           Paul Gouda/Se/Hydride Generation

Method Report
           Se/hydride generation/AAS

Atomizer: Flame
Matrix: ULTRAPURE WATER

METHOD INFORMATION **

Default Setup:

           Number of Repeats   : 1
           Flush Time (sec)   : 45.0
           Auto-Increment Sample Names? No

           Auto-print Calibration Curves?: Yes
           Analysis Graphics Display   :
           Absorbances
Auto-print Analysis Graphics? : Yes
Auto-save Analysis Graphics?   : No

Default File Names:

Analysis Data File : RESULTS
Sample Limits Table   : LCTAB
Autosampler Table: HYDRIDE
Blank (AZ) Limits Table : BLANKLCT

QC Check Table   : QCTAB

Recovery Check Table   : RQCTAB

OUTPUT INFORMATION
Output Mode: Concentration

Override Print Limits?   No
Override Signif Figs   No
Limits Table: LCTAB
Check?   No
Correction Factor: 1

Auto-print data?   Yes
Condensed report format?   Yes
Auto-store data?   Yes
Store individual repeats?   No

Report to:
    Screen   Avgs, Units
    Printer   Avgs, Units

Method: **PGSEHGA**   part-2
Paul Gouda / Se ? Hydride generation

Elements   : *Se*

Mode   : Double Beam Atomic Absorption

Integ Type : Automatic   Integ Time : 3.0 Sec.
Delay Time : 0.0 Sec.

Comments:

Flame Information:

Flame Type   : Air/Acetylene
Oxidant Flow (SCFH)   : 10.0
Oxidant Flow (SCFH)   : 4.5
Flame Type   : Air/Acetylene

Element   : Se
Element Name   : Se
Wavelength   : 196.00
Bandpass   : 2.0
High Voltage   : 700
BKG Mehtod   : None
Lamp Current   : 5.0
Signif Figs   : 4
Print Units   : ppb
Print Limits Low : 0
Print Limits High : 0
Stdzn Method: Multipoint Stdn.   Stdzn
Method: Standard Additions
    Std Names   Std Conc   Abs.
    Addn Name   Std Addn

A/Z#1: BlankSTD   = 0.00   = 0.0000
   #1: ADD1   = 10.00
   #2: ASSESB-1   = 1.00   = 0.0059
   #2: ADD2   = 20.00
A/C#3: ASSESB-5   = 5.00   = 0.0428
   #3: ADD3   = 30.00
   #4: ASSESB-10 = 10.00   = 0.0916

AAS Data documentation & communication "LIMS": Laboratory Information Management System:

a} Electronic :

1) Hard copy: C drive of the AAS, under "file: results", via "enable" (Illustration sample of common software).
   Hard disk back-up maintained regularly.

2) Software : Data transferred to a CD via "enable". Files are named to indicate element and date, e.g. Se042295.ASC. The data is edited after taking into account such factors as background correction, and the data is then transferred into the LIMS "network" system under "AA/data" {the DCI part of the reporting}

b) Conventional:

1) Project sheet: report indicating results; conc./ unit / dilution & volume ..etc

2) Computer printout: attach to the main project of the run {reference number}

3) AAS record binder, including performance record (e.g. calib. correlation coefficiency, SRM/QC record , production record , .. etc).

The Final chapter:

This book dealt with As / Se / Sb analysis within the scope of EPA methodology and range. It dealt solely - and as briefly as I could - with the areas related directly or significantly to the determination of these elements mainly via the hydride generation technique, graphite tube, flame and other Atomic absorption methods.

The sudden and positive change in the approach and attitude taken towards the environment in the past recent years necessitated the development of instrumental analytical techniques capable of determining concentration < 1 ppb. This is below what used to be reported just a few years ago as BDL {Below Detection Level} by conventional flame AAS at the ppm level.

The chemist will still run into samples with such high level that would necessitate bench or wet-chemistry dealing with g/l concentration and utilizting totration and precipitation techniques.

In fact, until a few years ago, reporting As / Se / Sb concentration in g/L was very common. Such gravimetric and volumetric wet methods can still be utilized with samples containing high level of As / Se / Sb as a preliminary or conclusive test, based on the determined BDL. Such unexpected level of contamination would require extremely high dilutions in order to apply ultra-trace methods such as ug/L - hydride generation. An example of such tests is the permanganate method and the potassium bromate titration.

The titration is based on the fact that:

1 ml N/10 KMNO$_4$ {or N/10 KBrO$_3$ } = 6.088 mg Sb

The principal of the potassium permanganate method is as follows:

2 KMnO$_4$ + 16 HCL + 5 SbCL$_3$ = 5 SbCL$_5$ + 2 KCL + 2 MnCL$_2$ + 8 H$_2$O

2 KMnO$_4$ + 16 HCL = 2KCL + 2 MnCL$_2$ + 5 CL$_2$ + 8 H$_2$O

The potassium bromate method is based on the fact that:

KBrO$_3$ + 3 SbCL$_3$ + 6 HCL = KBr + 3 SbCL$_5$ + 3 H$_2$O

Separation of selenium vis distillation as SeBr4 is another illustration:

H2SeO3 + 4HBr ⟶ SeBr4 + 3H2O

The old guideline of Se toxicity to humans is 5 ppm in food and 0.5 ppm in water or milk. In fact burns from hot acids containing Se as bromide show Se in urine. The author has experimented on and proven the effectiveness of bromobenzene in the body to eliminate selenium, as shown via urine analysis. Some forms of arsenic such hydrogen selenide are more toxic even at trace level it would cause eyes and nasal troubles. In fact all volatile selenium compounds including many of which you as a bench analyst will produce during sample treatment (in a fume hood, of course) such as heated dioxides, are dangerous.

Note that selenium dioxide is oxidized to S2O3 by many strong oxiding agents such as CLO3-, MnO4-, halogens, H2O2 and by electrolysis in acid and alkaline solutions.

Also note that these reactions are not quantitative unless the chemist forces them under arranged conditions beyond the topic of this chapter. Fusion with alkali metal carbonates, however, produces the salts of selenic quantitatively. Theoretically, aqua regia should produce these acids, however, you must keep in mind the very important reverse reaction that takes precedence:

$H_2SeO_4 + 2HCL \longrightarrow H_2SeO_3 + CL_2 + H_2O$

As you see, H2SeO4 is decomposed as soon as it was formed. CL2 doesn't oxidize H2SeO3 in an acidic solution because the HCL formed reduces H2SeO4.

Hydrogen sulfide will reduce the acids selenic and selenious as follows:

$H_2SeO_3 + 2H_2S \longrightarrow Se + 2S + 3H_2O$ (lemon yellow precipitation)
In the case of the higher-valence acids, the hydrogen sulfide first reduces it to the tetravalent sate:

$H_2SeO_4 + H_2S \longrightarrow H_2SeO_3 + S + H_2O$

Obviously, since selenic acid and sulfuric acid are isomorphous with one another, they will form salts whose crystals will mix in all proportions, e.g. CuSeO4.5H2O and CuSO4.5H2O

Se produces a vivid green solution when heated with concentrated sulfuric acid:

$Se + HeSO_4 \longrightarrow SeSO_2 + H_2O$

When diluted with water, red selenium is precipitated:

$SeSO_3 + H_2O \longrightarrow Se + H_2SO_4$

Iodometry

The iodometric method to determine Se is perhaps the best volumetric method if you are dealing with a sample such as industrial sludge or soil that has such level of Se. However, you must achieve a quantitative separation from other elements first. Ions such as Fe^{+3} and Cu^{+2} will also oxidize the iodide ion, releasing iodine.

Hence, to separate selenium from other elements, use $SnCL_2$ with distillation from a bromide solution. Traces of such elements as Fe or Cu occluded in the precipitated Se, can be removed by addition of phosphoric and tartaric acid for iron, or potassium cobalticyanide for copper. The bromine associated with selenium bromide in the distillation can then be removed by addition of urea, phenol or salicylic acid. Nitrous acid, which might (likely) be liberated by traces of nitric acid, can easily be removed by the addition of sulfamic acid or urea.

The most successful approaches to deal with the iodometric determination of selenium are:

- selenious acid is reduced with an excess of standard sodium thiosulfate solution as follows:

$$H_2SeO_3 + 4Na_2S_2O_3 + 4HCL \longrightarrow Na_2SeS_4O_6 + Na_2S_4O_6 + 4NaCL + 3H_2O$$

The excess $Na_2S_2O_3$ can then be back-titrated with standard iodine solution as follows:

$$2Na_2S_2O_3 + I_2 \longrightarrow Na_2S_4O_6 + 2NaI$$

1 ml 0.1N $Na_2S_2O_3$ = 1ml 0.1N iodine = 0.001974 g. Se. add 1 ml of 2% soluble starch soln. as the indicator for free iodine. Titrate at room temperature in 15% hydrochloric acid solution. The titration end-point turns colourless.

- A moderate excess of KI solution is added, and iodine is liberated as follows:

$$H_2SeO_3 + 4HCL + 4KI \longrightarrow Se + 2I_2 + 4KCL + 3H_2O$$

The liberated iodine in titrated against standard thiosulfate solution and in the presence of soluble starch. The equivalent factor is the same as the preceding titration. This titration is suitable for a higher Se level due to a less sharp (red selenium) end-point. Iodine tends to be volatile in high concentration solutions.

- The 3rd option is:
$$H_2SeO_4 + 2HCL \longrightarrow H_2SeO_2 + CL_2 + H_2O$$

Chlorine reacts with iodine as follows:

$$2KI + CL_2 \longrightarrow I_2 + 2KCl$$

liberated iodine is titrated the same way with standard sodium thiosulfate.

Note that the permanganate titration is free from interference by trace level ferric and copper ions. Nitric acid however would interfere with the back-titration of ferrous ion, and halogen acids would be oxidized by the permanganate ion. These acids need to be eliminated via gentle fuming with sulfuric acid in a large Erlenmeyer flask as follows:

$$2KMnO_4 + 5H_2SeO_3 + 3H_2SO_4 \longrightarrow 5H_2SeO_4 + 3H_2O + K_2SO_4$$

excess standard KMnO4 solution is back-titrated until the purple colour of permanganate ion disappears, using standard ferrous ammonium sulfate solution:

1 ml 0.1N KMnO4 = 1 ml 0.1N ferrous ammonium sulfate
= 0.003948 g Se.

The bench analyst needs to examine all possible and related chemical scenarios, complications and side-effects and must be able to approach each sample accordingly.

This applies to performing the bench - wet work as well as instrumentation. This necessitates:

- refreshing oneself with the theory part of wet chemistry.
- examining the methodology carefully.
- elucidating the intricacies of the topic through research literature.

The hydride generation technique and the use of graphite furnace atomic absorption and ICP, when applied correctly, is a very an acurate ultra trace method.

Recorded recovery of certified QC blind samples, spiked/Q3 samples, and Q2/standards attest to this fact.

About the author

Dr. Paul Gouda, C.Chem., P.R.MD., Ph.D.
Analytical Chemist.

- Dr. Paul Gouda, Ph.D. in analytical ultra trace analysis, and MD with a pharmaceutical research paper on chemical manipulation of neurological hormonal compounds.

- Former senior chemist at 4 major research analytical laboratories for over a decade.

- Has written several scientific papers and the books. His most recent book is a medical study in gender pre-selection "**Choosing the sex of your baby**: made simple, made certain." He has also produced a book similar to this paper, on Hg analysis "**Hydrargyrum, Hg analysis from diphenylthiocarbozone to cold vapor**."

   Tour his books at: **www.goudabooks.com**

- He has taught analytical inorganic chemistry at the college and the graduate levels.
- Chief chemist - consultant at Optimum Green Environmental Laboratories, the Canadian branch.

- Visit the author's chemical consulting & instructing site at: **www.optimumgreen.com**
- On a personal note, he is a single parent who currently resides in BC. Canada with his only son. His personal web site presents a variety of arenas and interests, including football "soccer" achievements to the highest level, poetry and psychology.
- The author can be reached at: **gouda@chemist.com**

www.ingramcontent.com/pod-product-compliance
Lightning Source LLC
Chambersburg PA
CBHW030948180526
45163CB00002B/710